Functional *Melodies*

Finding Mathematical Relationships in Music

Scott Beall

To DAD —
YOU ARE ULTIMATELY
RESPONSIBLE FOR THIS!
MUCH LOVE,
Scott

Key Curriculum Press
Innovators in Mathematics Education

Project Editor: Joan Lewis
Editorial Assistants: James Browne, Lara Wysong
Reviewers: Maureen Burkhart, Larry Copes
Production Editor: Kristin Ferraioli
Copy Editor: Mary Roybal
Editorial Production Manager: Debbie Cogan
Production and Manufacturing Manager: Diana Jean Parks
Production Coordinator: Jenny Somerville
Art and Design Coordinator: Caroline Ayres
Text Design and Composition: A-R Editions, Inc.
Text Design Graphics: Jason Luz, Jenny Somerville
Technical Art: A-R Editions, Inc.
Cover Design: Jenny Somerville
Cover Photo: © Tony Stone Images, Larry Ulrich
Prepress and Printing: Versa Press, Inc.

Executive Editor: Casey FitzSimons
Publisher: Steven Rasmussen

Key Curriculum Press
1150 65th Street
Emeryville, CA 94608
510-595-7000
editorial@keypress.com
http://www.keypress.com

Printed in the United States of America
10 9 8 7 6 5 4 3 2 1   04 03 02 01 00
ISBN 1-55953-378-1

Contents

A Note from the Author   **v**

Introduction   **vii**

How to Use This Book   **xiii**

Activity/Topic Chart   **xvii**

1   **Sound Shapes**   **1**
Hearing Geometry As Function and Metaphor
CD tracks 1–10

2   **Measures of Time, Part I**   **11**
Hearing, Writing, and Computing Fractions of Time
CD tracks 11–22

3   **Measures of Time, Part II**   **29**
Tempo and Rate Problems Facing Musicians
CD track 74

4   **The Multiples of Drummers**   **37**
The Mathematics of Polyrhythms
CD tracks 23–28

5   **Record-Producer Algebra**   **53**
Using Algebra to Perform Rap Music
CD tracks 29–34

6   **Functional Composer, First Movement**   **73**
A Mathematical Solution to Writer's Block
CD tracks 35–40

7   **Functional Composer, Second Movement**   **89**
The Relentless Composer
CD tracks 41–47

8   **Name That Function**   103
Determining Function Transformations by Listening
CD tracks 48–72

9   **Inside Out**   121
Hearing Pictures As Music Through Polar Coordinates
CD tracks 73–76

10   **Scaling the Scale, Part I**   143
Natural Vibrations and Pythagorean Tuning
CD tracks 77–88

11   **Scaling the Scale, Part II**   159
A Solution to the Limitations of Pythagorean Tuning
CD track 89

A Note from the Author

One day while substitute teaching a middle-school mathematics class in Daly City, California, I encountered the ultimately uncontrollable classroom. With one week of classroom experience under my belt and no teacher training, I wondered, what could I do to deal with this chaos? What might possibly get through to these kids? Many of my students viewed mathematics as dull, difficult, and boring—something to be avoided at all costs. However, I also found these same students glued to their CD players for hours each day—they held music as their greatest ally and the ultimate expression of their identity. Intrigued, I ran home at lunch to get my programmable electronic drum machine. I decided to try to make some mathematical sense out of what was already such a big part of my students' world: music. I found it worked.

That day in Daly City stuck with me, and clearly I wasn't alone in my thinking. In the years that followed I noticed that people often commented on how mathematical music is. When I looked for more information on the topic, I found volumes of literature on the music and mathematics connection. I discovered that this connection has fascinated great thinkers and entire cultures for millennia, starting in recorded history with the Pythagoreans. But I also discovered that the music and mathematics connection was all but absent from today's school curricula. Was this an oversight on the part of educators? When I reflected on the connection between music and mathematics, the educational opportunities it presented were impossible to deny. Music seemed like a key for reaching students in mathematics. My inspiration was sparked.

Too much of what I love about mathematics is left untouched in classrooms. Rarely are students encouraged to wonder about the role mathematics plays in the larger contexts of nature, culture, and philosophy. Too often the experience of learning mathematics is restricted to that of rote memory and calculation. Thus it is no wonder that mathematics appears irrelevant to many students—ways for them to apply meaning and imagination to what they learn are scarcely explored. Student failures in mathematics are today at crisis levels. Is this a coincidence? Hardly. I became further inspired to improve mathematics education.

This book is the result of that inspiration and much thought, research, development, and testing in the classroom. Though it merely scratches the surface of how the music and mathematics connection can be used in the classroom, it is my hope that it will help to expand the way you and your students think about teaching and learning mathematics.

Most important, it is your excitement and inspiration that will translate to students. The most advanced curriculum and technology is no match for the passion of teachers sharing what it is about mathematics that they love and teaching what it is about mathematics that inspires the imagination.

Acknowledgments

Heartfelt thanks and appreciation go to:

My father, late mother, brother, and sister, whose love, inspiration, and collective mathematics and music expertise influenced my vision and insights for the many connections made in this book. Special thanks to my sister Kris Berman for providing valuable advice on parts of the manuscript.

The students and teachers at Homestead High School, Cupertino, California, who tested the activities and offered encouragement, helpful criticism, and support.

The faculty of the Stanford University Teacher Education Program, particularly Beverly Carter, Vicky Webber, Joanne Lieberman, and Gary Tsuruda, for their encouragement and vital critical feedback in the early stages.

Thomas Milke for his patience and engineering expertise in recording the CD, and Ed McClary for providing McClary Music Studios.

The staff at Key Curriculum Press for their belief in the project that allowed it to become a reality.

All of those in my professional life and elsewhere who expressed excitement about the potential of teaching mathematics with music, and encouraged me to "write the book."

Scott Beall

Introduction

Functional Melodies is intended to be taught as a supplement to a primary mathematics curriculum. Students use the activities to actively experience the integration of music and mathematics by listening to and performing music as they review mathematical concepts, solve problems, and perform calculations. For the most part, the activities are independent of each other, have no musical prerequisites, and can be used in any order.

Read through this introduction before starting the activities. It will explain the nature, history, and pedagogy of the music and mathematics connection and provide specific instructions for using this book. You will be best equipped to adapt and modify the activities to suit your needs and style if you understand the music and mathematics connection, as well as what outcomes to expect for student learning and classroom culture. Given that you are reading this book, you already are interested in its potential. Let your excitement show in the classroom. Your enthusiasm, confidence, and inspiration around the connections that the activities make can be as important for your students' success as any logistical preparation.

Care has been taken to make the materials friendly for all mathematics teachers regardless of their musical background. The audio tracks on the accompanying *Functional Melodies* CD allow you to conduct most of the activities without the use of musical instruments; all that you need is a CD player. There are, however, opportunities for you and your students to use musical instruments in place of the CD, turning your mathematics class into a music studio.

The Music and Mathematics Connection

The complex connection of music and mathematics exists on many levels, from the concrete to the abstract. The activities in *Functional Melodies* demonstrate the music and mathematics connection on four of these levels: the physics of sound, musical language, aesthetics, and metaphor.

Physics of sound   Frequency, waves, resonance, vibration, and the mechanics of how musical instruments create pitch are all easily modeled by mathematics. Activities 10 and 11 (Scaling the Scale, Parts I and II) deal directly with the physics of sound.

Musical language   The systematic organization of pitch and rhythm for the creation of music can be thought of as the mechanics of musical language. Pitch-interval relationships that create scales and chords, harmonic relationships, the division of time into patterns of rhythm, and musical form all contain elements that relate to each other by a set of quantitative rules with their own musical symbols. This connection is still at the mechanical level, and although it does imply rules of aesthetics and art, it is not concerned with that dimension. Activities 2 through 4 (Measures of Time, Parts I and II, and The Multiples of Drummers) deal with pitch and rhythm relationships.

Aesthetics   Aesthetics is an intriguing level of analysis where the role of mathematics lies between the practical and the metaphorical. Mathematical analysis lends insight into aesthetically pleasing harmonic, melodic, and rhythmic patterns in Activity 6 (Functional Composer) and Activity 9 (Inside Out).

Metaphor   The metaphorical connection between music and mathematics is largely based in process and analogy. The process of composing music can be likened to mathematical problem solving, and many musical systems and structures can be viewed as analogous to physical structures that can be modeled by mathematics. Activity 1 (Sound Shapes) is a direct example of this. Activity 9 (Inside Out) also treads in this area.

Connections Through History

For the Pythagoreans of ancient Greece (approximately 500 BC), the study of numbers and how they relate to musical harmony was considered the path to reaching spiritual understanding and purity of soul. Together, music and mathematics provided keys to the secrets of the world. The doctrine of the "music of the spheres" held that human souls must be attuned to the laws of the universe and suggested that planets in space produce sounds, a singing universe. (Later, Plato clarified this concept in his writings of *The Republic* and *Timaeus.*) The *Quadrivium*, the highest curriculum studied by the Pythagoreans, included music, geometry, arithmetic, and astronomy. Music was studied entirely as a mathematical medium. In fact, music was considered so purely mathematical that there is no record of its aesthetic dimension at this time, and there is virtually no record of folk music from early Greek culture.

Since the Pythagoreans, there have been many musician-mathematicians. Nicomachus (AD 100), Ptolemy (AD 165), Boethius (AD 500), Kepler (1600), Mersenne (1600), and Bernoulli (1700) all published works and led their own music and mathematics movements. Particularly fascinating is the work of Joseph Schillinger in 1945. The *Schillinger System of Musical Composition* was a massive undertaking employing a multitude of mathematical structures and formulas to aid in the refinement of musical composition by solving musical problems and communicating and classifying musical ideas. George Gershwin, Schillinger's best student, used this system to a significant extent in the opera *Porgy and Bess.*

The past several decades have seen a revolution in the way music is performed by many popular artists and used by the media in advertising, television, and the movies. The advent of digital sampling technology, computer sequencing, and MIDI (Musical Instrument Digital Interface) capabilities have often reduced the performance element of music to a process of computer programming and MIDI configurations. Many of the instruments heard in popular music and advertising are computer performances of digitally sampled sounds. Music is even being composed by computers. At Stanford University, the Center for Computer Research in Music and Acoustics (CCRMA) is pioneering research in the application of mathematical algorithms to musical composition.

Some may find the idea of the mathematization of music performance and composition a bit disturbing. Though today's high school students have grown up with computer-generated music, they often express pointed concerns about the mathematization of art and aesthetics. The activities of *Functional Melodies* can increase your students' awareness about the many ways mathematics can relate to aesthetics.

Art, Aesthetics, and the Domain of Mathematics

Why do we find a piece of music pleasing? What distinguishes random noise from music that captures our ear? How does a musical composition work? What holds the parts together? And, what does mathematics have to do with any of this?

These questions on the music and mathematics connection can lead to thought-provoking discussions—many of which are raised in the activities of *Functional Melodies.*

The creative process Musical composers draw from two primary domains in their work: inspiration and technique. Sometimes composers do not think of anything specific during the composition process; the music virtually appears to them, and all they have to do is write it down. This type of music comes from pure inspiration with no conscious application of theory or technique. On the other hand, when they lack inspiration, composers can use various formulaic methods to think through the creative process and compose their music. In practice, most artistic efforts are a blend of these two extremes. A bout of inspiration comes along and the composer's theoretical understanding and technique allow him or her to realize and articulate the message. The creative process swings between conscious application of discipline (rational, mathematical) and pure inspiration (intangible, spiritual). In the end, any aesthetically pleasing music contains some structure that binds it together and gives it a coherence that communicates. The search for this unifying structure of aesthetics leads directly to mathematics.

The role of mathematics Our discussion suggests that successful music contains a mathematically unifying structure. Is the converse true? Does the presence of mathematical integrity guarantee the success of a piece of music? Can the elusive artistic elements of inspiration and "chemistry" in music be measured or modeled by mathematics? Pythagoras asserted that numbers held the potential to explain all of reality. But as any experienced musician knows, a successful performance involves more than technical or mathematical correctness.

An intriguing idea to consider for classroom discussion is how the advent of the new mathematics of chaos and fractals might be able to probe the subtleties of music performance. Infinitely iterating models may be able to provide mathematical insights into realms commonly perceived as intangible. These can be potent topics for students. But be forewarned: I have had some students become very upset at the notion of using mathematics to analyze music on the emotional level. I have had to reassure them that there will always be domains beyond the mathematically measurable. Through these discussions, I have also discovered attitudes in my students that I never knew existed and I have gained some insights into hidden effects of mathematics schooling in general. A surprising number of students have very distinct and sometimes limited ideas about what mathematics can and should be used to explain. This is cause for wonder on the part of educators—to what extent might some mathematics curriculum and instruction stifle creativity and open-mindedness in our students?

Teaching and Learning with *Functional Melodies*

To use this book most effectively, it is important to consider some of the pedagogical premises on which it is based.

Mathematics as the story, music as the language   The work of Howard Gardner on multiple intelligences suggests that for education to address the needs of all students, information must be made available to them in many forms. If a story needs to be told but is written in a language that the audience does not understand, the story must be translated to the appropriate language to be comprehended. In *Functional Melodies,* mathematics is the story and music is the language that makes it comprehensible. Students perceive and measure quantitative relationships (mathematics) through sound (pitch) and body interaction (rhythm). Many people can more readily hear the nature of a quantitative relationship or pattern than they can calculate it numerically or perceive it visually. In the two Functional Composer activities, the quantitative relationships of graph transformations are expressed aurally, visually, and numerically, so that the nature of the relationships are accessible to a broader audience, and can be understood more deeply at their mathematical level.

Depth of understanding through varied contexts   How do we really get to know a concept—or our best friend, for that matter? Deep understanding is gained by experiencing the subject in a variety of contexts and viewing it from many perspectives. We get to know people better when we share experiences with them in unique situations, such as traveling to a foreign country. Getting to know mathematics can happen in a similar way, and doing mathematics in a musical context can reveal the essence of a mathematical principle from a unique, and otherwise unobtainable, perspective. In my personal experience with students and teachers, I often hear comments such as, "I never *really* understood functions until I saw this. . . . "

Balancing the classroom culture   Much research has been done on the social construction of intelligence in classrooms and its effect on learning. Any experienced teacher has witnessed how a pecking order of "smart kids" versus "slow kids" can evolve between students. This can polarize the learning environment and program certain students to failure. Astute teachers aware of this seek out methods to establish a classroom culture that overcomes students' perceptions of ability. When a struggling mathematics student who is musically inclined suddenly becomes the expert in a class activity and is sought out for help by a "smart kid," an invaluable shift takes place in the classroom status culture. Students' confidence is enhanced and the learning environment is made safer.

Cooperative skill building   Most of the activities in *Functional Melodies* require interdependence between a pair or group of students. This design of intentional interdependence maximizes student interaction. Students learn the efficiency of working with peers.

Enhancing appreciation for the scope of mathematics   Interdisciplinary connections enhance a sense of meaning that motivates students to learn. When a subject such as mathematics is connected to an integral part of adolescents' culture such as music, mathematics is made fun and revealed to be a part of their worlds in a way they never imagined. While isolating disciplines can be instructionally efficient in many instances, it often leads to a sense of irrelevance and meaninglessness. Integrating disciplines reconstitutes the world, returning it to its authentic state—whole, interactive, and interdependent.

Learning Pathways

Learning takes place in many ways in heterogeneous classrooms, depending on the type of instruction practiced by teachers and the structure and content of curriculum. To effectively evaluate curriculum and instruction, it is important to realize that learning occurs both directly (immediately) and indirectly (delayed). In the case of direct learning, students acquire information, insights, and skills directly from the teacher or through curricular activities. The outcomes of direct learning are clear and assessable immediately following the instruction. Indirect learning, a more circuitous route of student growth, takes place as a more self-paced learning process within the student. With indirect learning, the ultimate outcome may not be predictable nor can it be immediately evaluated. Curricular activities that promote indirect learning inspire curiosity and imagination, supply a connection lending meaning to another topic, or provide a ground for future experiences to be more fully understood. Indirect learning is a part of the constructivist learning model, where curriculum and instruction facilitate students to construct their own meanings and secure a solid understanding of the material.

Functional Melodies activities may impact students both directly and indirectly. The flowchart below presents a model of how this can happen. Each bubble in the flowchart indicates a different outcome a student may experience directly from an activity. At this point an indirect learning process may take over, leading the student through the other outcomes in the flowchart, and ultimately to achieving mathematical skill and understanding.

Learning Pathways in *Functional Melodies*

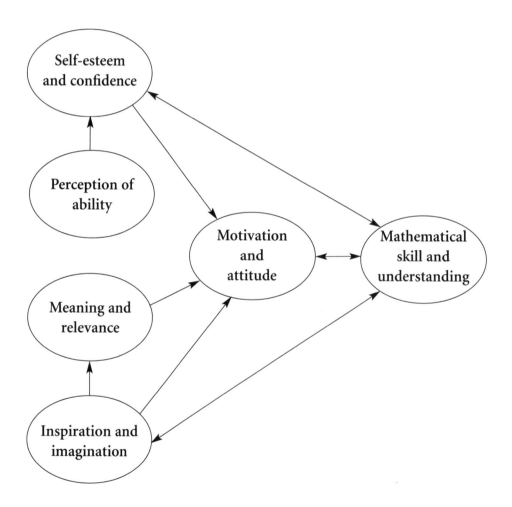

How to Use This Book

Choosing an Activity

You may want to select an activity as enrichment to support and align with the mathematical content you are currently teaching. Or you may want to choose an activity solely to shift instructional mode, to stimulate individual students, or to energize the classroom culture.

Aligning with content   The teacher notes for each activity indicate how it can be connected to the primary curriculum and what mathematical topics it contains. Survey your course topics for the year and refer to the *Functional Melodies Activity/Topic Chart* on page xvii to find appropriate fits. It is also important to keep an open mind about ways to adapt the activities. For example, I have used very effective adaptations of Functional Composer with third-grade students. Don't rule out an activity if on face value it contains topics that are associated with a different mathematics level or not aligned with your course. Some suggestions for adapting the activities are provided in the teacher notes.

Shifting the instructional mode   Students stay more engaged when the mode of instruction is varied. Since these activities stand alone with few prerequisites, it is possible to use them at any time. Between units and before and after school vacations are great junctures to use these activities. However, when concrete relevance of the activity to the course content is unclear, students can be resistant. If you use the activities out of context with the primary curriculum be aware of your students' sensibilities and provide adequate background preparation with the readings and discussion.

Instructional Format

Functional Melodies activities place you, the teacher, into the learning experience together with your students. Activities usually begin with you facilitating an exploration using the accompanying music CD and engaging the entire class with questions. The focus then oscillates between directed discussion by you and students working in pairs or groups to complete tasks or explorations on their worksheets. In some ways, it can be an advantage if you have no musical background! This authentically places you in the role of learner with the students. You can let go of your ego here—it is completely valid for a mathematics teacher to be ignorant of music theory. Consider it an opportunity to model the ethic of lifelong learning. It can be great to have your musical students be the ones to clarify any confusion that may result from lack of music knowledge within the class.

Student Materials

Worksheets   Even though they work in pairs or groups for most of the activities, all students should record their own work on individual activity worksheets. The worksheets vary in style; some are graphs and charts while others contain problem sets that the students complete individually, possibly as homework. Make copies for each student from the blackline masters provided.

Resource pages   Resource pages are a source of information for student groups. They contain summarized instructions of the goal of the activity and explain what students are expected to do, making them less dependent on direct instruction from you. In some cases, the resource pages contain necessary information such as vocabulary, definitions, charts, graphs, or diagrams. Each work group should have one resource page. The objective here is not merely to save paper. One sheet per group fosters interaction and interdependence among group members. Students will have to read the resource page together, or if one student reads it, he or she will inevitably be asked by the others to explain it. While I recommend that you use the resource pages, you can make available the information they contain in any way that fits your style. You might want to display one as an overhead, copy some material to the board, or give the information in a direct-instruction lecture format. Keep in mind that the resource pages are meant to create a comfort level for the students, to foster interaction between group members, and to reduce students' dependence on you.

Readings   For most students, the idea that music and mathematics are closely related is a foreign concept. Thus there may be resistance or critical attitudes from students when relatively untraditional activities are presented. The *Functional Melodies* readings are designed to set the context for students and to enhance their readiness to engage in the activity. The readings are purposefully light and conversational to set an inviting and friendly tone. They are intended to inspire imagination and curiosity about the activity topic, to raise essential questions, to orient students to the concepts and issues presented in the activity, to introduce general background information for relevance, and to provide specific background knowledge necessary for the activity.

There are several ways to use each of the readings: Give a copy to each student a day or so prior to the activity to read for homework, ask a student to read it to the class or read it to the class yourself, or summarize the information in the reading and present it to students as a lecture or skit. No matter how you present the reading, discuss the ideas it contains either before the day of the activity or on that day.

Teacher Materials

Audio recordings   Track numbers from the *Functional Melodies* CD are clearly indicated in the teacher notes for each activity. A CD player with a pause feature and ability to recall specific track numbers will make it easy to play the correct track. Some of the activities require you to repeat a particular track several times. You may also choose to use live performances by students (or yourself) if you wish.

Overhead transparencies   Transparency masters are provided for Activities 8 and 9. For the activities with graphing exercises, you will find it helpful to make overhead transparencies of the activity worksheets. In Activity 11, you may want to make a transparency of the resource page.

Activity scripts   The teacher notes for each activity contain a step-by-step script to guide you through the lesson. Included in these steps is the information essential for conducting the activity, such as questions for students and issues that may arise. Where the script suggests specific questions to pose to the class, the answers are noted in square brackets directly following the question. Other features presented in the activity script are discussion points and teaching tips.

 Discussion points   Discussion points indicate opportunities where you may wish to pause for discussion and elaborate on a relevant connection. They are interspersed throughout the activity scripts and indicated with the icon shown at left. A discussion point may be a series of questions or a presentation of interesting facts that relate to issues the activity raises at that point. Some cases warrant taking a significant digression from the activity to present examples from the primary curriculum that demonstrate the connection to music that the activity is exposing. However, discussion points are completely optional. All of the activities can proceed successfully if no discussion points are used.

Discussion points may contain interdisciplinary connections that are historical and cultural, special interest items, and specific connections to the primary mathematics curriculum. They may also provide educational opportunities for topics that vary from specified connections to the primary curriculum.

As you decide which discussion points to use, consider the following: your own background and comfort level; student background, skill/ability level, and maturity; time allocated for the lesson; overall curricular objectives; and tone of the classroom at any point in time. As you read through the activity script, become familiar with all the discussion points and tag those that you might use, but remain open to the possibility of using any of them. The first time you use an activity, it is difficult to predict what students will ask or what the classroom climate will be.

Teaching tips Teaching tips differ from discussion points in that they offer advice on issues of pedagogy and instructional strategy and do not contain material to be used directly with the students. Teaching tips are indicated in the activity script with the icon shown at left. They contain general pedagogical issues, insights into student issues with a particular aspect of an activity, background information on a topic that may assist you in conducting the activity, and my own anecdotal experience and advice.

Follow-up activities At the end of each activity, suggestions are provided that connect the ideas to the primary curriculum or allow for deeper study and extension projects.

Each activity of *Functional Melodies* is a starting point for a different music and mathematics connection. In writing this book I have barely scratched the surface. I hope that you and your students can uncover some new connections through research and your own creativity that will inspire and enrich your experience of mathematics. Good luck!

Activity/Topic Chart

Activity	Activity Mode	Mathematics Topics	Music Topics
1 Sound Shapes	Listening, sketching, discussing	Geometry, assorted vocabulary terms and concepts, mathematics as metaphor	Ear training, form and composition, melodic structure, rhythmic patterns
2 Measures of Time, Part I	Listening, creating models, singing, calculating	Measurement, addition of fractions, symbolic representation	Rhythmic notation measures, time values of notes, beats
3 Measures of Time, Part II	Listening (minimal), paper and pencil calculation, problem solving	Rates, unit conversions, dimensional analysis, operations on fractions	Rhythmic notation, tempo
4 The Multiples of Drummers	Interactive, listening, ensemble performance, calculating	Counting, multiples, least common multiples, factoring, exponents, ratio, patterns, problem solving	Beats, tempo, polyrhythms, phrasing, rhythm, accents, performance
5 Record-Producer Algebra	Interactive, listening, vocal performance, problem solving, manipulation of symbols	Counting, creating algebraic representations, solving equations	Beats, time, measures, pop/rock song arranging, vocal phrasing, rap vocal performance
6 Functional Composer, First Movement	Interactive, listening, identifying and calculating, graphing	Definition of a function, function notation, graphing transformations of functions, composite functions, order of operations, in/out tables	Musical notation, music composition and melodic structure, ear training, transposition, modes
7 Functional Composer, Second Movement	Interactive, listening, graphing, calculating	Graphing transformations of functions, periodic functions, in/out tables, function notation, domain, range	Melodic structure, patterns, intervals
8 Name That Function	Interactive, listening, singing, calculating	Function notation, transformations of functions, calculating function values, problem solving	Pitch, melody, ear training
9 Inside Out	Interactive, composing, drawing, tracing, listening, performing, graphing	Polar coordinate graphing, scaling of axes, angle measurement, angular velocity, mathematical interfaces	Musical notation, staffs, melodic contour, composition, imagery in music, instrumental performance
10 Scaling the Scale, Part I	Interactive, listening, calculating, problem solving	Ratio, operations on fractions, problem solving, the work of Pythagoras, physics of sound	Scales, intervals, pitch recognition, frequency of pitch, harmonic series, piano keyboard
11 Scaling the Scale, Part II	Calculating, problem solving	Geometric sequence, patterns, multiples, fraction multiplication, ratio, problem solving	Scale temperaments, chromatic tones, intervals, piano keyboard

Sound Shapes

Hearing Geometry As Function and Metaphor

Sound Shapes explores various elements of music that embody geometric principles. The introductory reading discusses how the study of mathematics can influence our lives beyond its direct utilitarian applications by enhancing analytic thinking and enriching communication through metaphor. This activity presents music as an example of the use of mathematical terms as metaphors to communicate nonmathematical ideas. Students review a list of geometric terms and refine and agree on their meanings. They then listen to ten music examples, each of which illustrates a geometric concept in its musical structure. Students analyze the music and match it with the geometric term that denotes a relationship between geometric elements analogous to the relationship they hear in the notes or rhythm. Sound Shapes does not require students to do mathematical calculations. It uses indirect learning pathways to enhance understanding of geometric terms.

Mathematics topics

Geometry vocabulary, mathematics as metaphor, symbolic representation, the use of mathematics in daily life. *Prerequisites:* An understanding of the meaning of geometric terms.

Music topics

Ear training, melodic voice leading, melodic shape. *No prerequisites.*

Use with the primary curriculum

- As enrichment when you are introducing definitions of mathematical terms
  Use Sound Shapes when you are introducing new terminology. It need not be confined to geometry class, because the message of the activity reaches beyond the specific content of the terms.

- As a special interest activity at any juncture in your curriculum
  Sound Shapes can provide a thought-provoking segue between units in any course.

Objectives

- To enhance understanding and retention of geometric concepts
  The unique application of geometric terms in Sound Shapes requires students to fully understand the underlying ideas.

- To inspire students to study mathematics
  Sound Shapes shows how the study of mathematics can be used to provide metaphors for nonmathematical concepts. The activity can inspire new ways of perceiving the value of mathematics and the scope of its applications.

- To teach abstraction and metaphor
  Sound Shapes stimulates higher-order thinking. Developing students' abilities to think abstractly and to use metaphor can prepare them for advanced mathematics.

- To provide access for auditory learners
  Sound Shapes allows auditory learners to shine, while usually strong students may struggle. This can be a healthy shift in classroom culture, with benefits that can last throughout the school year.

- To stimulate students by providing pedagogical variety
  Sound Shapes provides a refreshing change of instructional pace that can stimulate students and carry over to activities that students find less engaging.

Student handouts

- The Hidden Life of Geometry (reading; one per student)
- Help, Hints, and Geometric Terms (resource page; one per group)
- The Geometry/Music Connection (worksheet; one per student)

Materials

- CD tracks 1–10

Instructional time

50 minutes

Instructional format

Students work in groups of up to four. Since most classes demonstrate a large diversity of student ability to think in the abstract ways that Sound Shapes requires, make the groups as heterogeneous as possible in all respects, including musical background, mathematical strength, gender, ethnicity, language proficiency, personality type, and learning style. You may be surprised by which students are able to make the abstract connections.

Following the reading, you can either lead the initial discussion or assign a student to facilitate it. Your role during most of the activity will be to play the CD tracks, coach students as they work in groups, and lead discussions. For each of the ten examples, the activity focus moves from independent work groups to whole-class discussion. The structure of the activity is quite simple. The complexity lies in developing a level of comfort with its abstract nature and effectively coaching students.

Student preparation

Have students read The Hidden Life of Geometry the night before or the day of the activity. If it is the norm for your class to engage deeply in group discussions, you might present the reading the day before and have an opening discussion at that time. The activity itself will take 30–40 minutes.

ACTIVITY SCRIPT

STEP 1   Discuss the topics of the reading

This is a good opportunity to have an open discussion on a variety of issues in mathematics education.

Ask students:

Why is it so important to study mathematics in school?

In what ways does mathematics prepare us for life?

How can knowing mathematics help us communicate more effectively about nonmathematical things?

What is a metaphor?

What are *parallel lives* and what does *circular reasoning* mean?

What do you think the reading was talking about when it said that mathematics can help us with our romantic life? Is this idea far-fetched? [Mathematics develops skill in logical/rational thinking and systematic problem solving. If we need to make decisions when we are emotionally distracted by love or tragedy, training in rational thinking can be a skill to fall back on and can help us find clarity.]

Give students an overview of the activity and direct them to refer to the resource page Help, Hints, and Geometric Terms.

STEP 2 Review the definitions of the geometric terms

Systematically go over the geometric terms on the resource page to reach clear mathematical definitions that use exact language. This important step will help students make the musical connections. For example, we might first describe the term *intersection* as "two lines crossing," then refine this definition to "the point at which two lines cross." To use the idea of intersection in a broader context, such as the description of a musical event, we can then define it as "the set of elements common to two sets." In the first music example, two instruments are playing melodies. These two melodies each comprise a set of notes. At one point the two instruments play the same note, creating a common element for the two sets. This note represents the intersection of the two melodies.

As you coach students to make the connections between geometric terms and music examples, revisit and rearticulate the meanings of the terms.

STEP 3 Discuss how to listen to the examples

Before beginning the listening exercises, direct students' attention to the resource page for guidelines for listening to the music. Be sure they understand the idea that melodies have a shape that can be thought of as a graph. (See Functional Composer [both movements] for specific applications of this concept.)

STEP 4 Work through each music example, matching geometric terms

Use the teacher notes in the Answer key to coach students and verify responses as they work through the examples. Play each example as many times as necessary. For each example, have groups discuss among themselves to arrive at a set of choices before you switch to class discussion. When the whole class has discussed the rationale for the primary choice and the second and third options, move on to the next example.

For each example, students make a sketch illustrating the primary geometric idea and write a statement explaining its connection to the musical example. There is no single, absolute right answer for each match, but some answers are more justifiable than others.

 While the choices for the geometric terms have a subjective/interpretive aspect, it is important to insist on a firm rationale for the way the term is embodied in the music example. These justifications can lead to many interesting and thought-provoking class discussions. Keep in mind that one objective of the activity is to draw in students who may not ordinarily participate by stimulating imagination and opening discussion.

STEP 5   Writing

Remind students of the discussion with which you opened the activity and use the questions from Step 1 to generate a short essay/reflection on the activity. Suggest that they refer to ideas raised in The Hidden Life of Geometry. It is important that students write while the experience is fresh, either on the day of the activity or on the same night as a homework assignment. A written reflection after an activity of this type is extremely valuable because it captures the insights and awareness gained and integrates them into class culture and subsequent academic work.

FOLLOW-UP ACTIVITIES

Textbook assignments

Sound Shapes can provide a rich and thought-provoking backdrop to textbook assignments that teach definitions of mathematical terminology.

Projects

- Carry out a scavenger hunt by searching through magazines, newspapers, TV shows, and any other form of media to find examples of mathematical terms being used as a metaphor for something nonmathematical. Explain how the metaphors work, and present the findings to the class.

- Write a poem, essay, or short story that uses the geometric terms of Sound Shapes metaphorically.

ANSWERS

The Geometry/Music Connection

CD track	Primary geometric term	Other reasonable terms	Written description and notes
1	Intersection	Oblique Symmetric Angular Reflection	Students may choose *oblique* and *angular* before *intersection*. The piano plays musical lines that cross each other and share one note in common. The common note is a little difficult to hear. If the notes of each melody comprise a set, then the notes (elements of the set) in common form the intersection of the sets. The intersection in this case is one note.
2	Oblique	Angular	One instrument repeats the same note while the other moves. This example also suggests the image of an angle. Musicians call this *oblique motion*.
3	Parallel		*Hint:* Listen to the relationship between the notes of the two instruments. The notes that the two instruments play stay the same "distance" apart from each other for the extent of the musical line. Composers use the musical term *parallel motion* in melodic lines and pay close attention to this important element. It has the musical effect of diminishing the listener's ability to perceive two melodies as distinctly different.
4	Even	Symmetric Periodic	The primary rhythm cycle is an even number, 2 or 4 depending on how you count. Suggest to students that they count with the music and look for patterns of groupings of accents. Jazz and rock musicians commonly characterize a rhythm as *even* or *odd*.
5	Odd	Symmetric Periodic	The primary rhythm cycle is an odd number, 7, or can be counted as 3 and 4. This example is difficult.
6	Angular	Oblique Square Obtuse	If graphed, the contour of the melody would look like a jagged mountain range—sawtooth, very angular.
7	Square	Symmetric Periodic	This example illustrates a metaphorical and abstract way to describe melodic shape. Unlike terms such as *parallel* and *oblique*, which have specific technical definitions in music, a descriptor such as *square* is more subjective, describing character or feeling.
8	Congruent	Reflection	The two melodies are exactly the same in size and shape, hence congruent. (If performed at the same time, the musical term would be *unison*.) *Reflection* is not mathematically or musically correct, but it is intuitively logical.
9	Reflection	Symmetric Similar Congruent	The second melody is a vertical reflection around the beginning note. When one melody goes up, the other goes down by the exact same amount, like a mountain range reflecting on a lake. This is the same relationship as that of the graph of $y = f(x)$ to $y = -f(x)$. Composers have used techniques like this to compose melodies for hundreds of years. They are the topic of the activity Functional Composer. (Musicians use the term *inversion*.)
10	Round	Circular Stretched Shrunk	This is another subjective, metaphorical reference.

THE HIDDEN LIFE OF GEOMETRY

Did you ever wonder why you have to learn all those vocabulary terms in geometry class? After all, what do they have to do with anything that matters in real life? How do they relate to anything outside geometry class?

If you ask your teachers, they will tell you that knowing mathematics and geometry is important in countless professions. At the very least you need to take math classes in order to graduate from high school, and if you want to get into the college of your choice, you will need good math grades. But if you aren't going into a job that uses the mathematics you are learning and you don't plan to go to college, it can seem as if you are learning math just to jump through hoops to get to another place in your life.

As you study mathematics in school year after year, trying to master each topic and stay on top of things, you can sometimes miss why, where, and how it is helpful. No matter who you are or what you do, whether you're a convenience store clerk, a student, an artist, or an astronaut, you are always making decisions about all kinds of things: what to eat for dinner, what clothing to buy, who to hang out with on the weekend, what streets to take to get downtown, and so on. Sometimes you might have difficulty making decisions and must then stop to "think it through" to figure out what decision will get you what you want. You analyze the situation and apply logical thought. Believe it or not, one big reason for studying mathematics is to do this very thing: to strengthen your ability to think logically and analyze things—even things that don't seem to be particularly mathematical. The ability to think logically can even be helpful when you are confused in your romantic life!

There's another side to how mathematics can enhance daily life. No matter what we do, most of us would say we want to be taken seriously by the people around us and be convincing when we communicate. In ways that might surprise you, studying the concepts of mathematics can help you with just that—expressing yourself better when you talk about things that have nothing to do with math! Many of the mathematics ideas you learn have counterparts in the real world that you might not have thought of. Geometric terms can be especially powerful in communicating ideas. People speak of having *parallel lives*. What does that mean? A conversation or argument may be called *circular*. What does that mean? These are examples of using a mathematics term as a *metaphor* for something else. A metaphor compares and connects two ideas that are essentially different but share some aspects on a meaningful level.

In Sound Shapes you are going to look at an example of how this idea applies to geometry and music. Musicians use mathematics ideas frequently, sometimes as metaphors and sometimes more directly. In order to keep track of notes and their relationships, musicians often create geometric mental pictures of the notes. In fact, many geometric terms and ideas are musical terms as well. As it turns out, musicians often think the same way as mathematicians and use geometric ideas to describe and organize what music does.

Music is just one example of using mathematics ideas to express nonmathematical things. In Sound Shapes you will explore these connections.

HELP, HINTS, AND GEOMETRIC TERMS

Understand the geometric terms.

Survey the list of terms and take notes from the class discussion on the exact meaning of the terms. Be careful not to assume that you know everything about what the term means. You need to know the meaning on a deep level to make a connection to music.

Listen to the music.

Your teacher will play an assortment of music examples. You will hear melodies, rhythms, and other musical patterns. Here are some questions to consider with your group to help you match the music to the geometric terms:

- Melodies
  How do the melodies relate to each other? Are they both changing? Do they change the same way together? Do they cross each other? How does their distance from each other change? Do the melodies suggest some kind of shape? Imagine that the pitch of the melody could be graphed, with high pitches high on the graph and low pitches low on the graph. Try to visualize what the melody would look like.

- Rhythms
  Try counting with the music. How are the beats grouped together?

Match the terms to the music.

Look at the list of geometric terms and find the term that best describes what is happening in each music example. If you get stuck, review the exact definition of each term and see if any aspect of the music example illustrates that idea.

Make alternate choices.

Several terms may fit a particular music example. Write these terms on the Geometry/Music Connection worksheet and explain what aspect of their meaning is demonstrated by the music.

Geometric terms

translated	angular	stretched	supplement
oblique	congruent	shrunk	adjacent
square	periodic	similar	even
round	reflection	intersection	obtuse
straight	symmetric	odd	
circular	parallel	space	

THE GEOMETRY/MUSIC CONNECTION

Music example	Geometric term	Sketch (Draw a sketch to show the meaning of the geometric term.)	Written description (Explain how the geometric term applies to the music example. Use the exact definition of the geometric term.)
1			
2			
3			
4			
5			
6			
7			
8			
9			
10			

2

Measures of Time, Part I

Hearing, Writing, and Computing
Fractions of Time

Measures of Time, Part I focuses on rhythm and shows that it, like mathematics, is a universal form of communication. In the activity, students use fraction relationships to describe relative durations of sound. The quantities being fractionalized are purely auditory, so a symbolic means to represent what is heard is developed through a directed discovery process. Coached by the teacher and the audio tracks from the CD, students subdivide a given time interval into primary units (measures), then subdivide the measures into even-multiple subdivisions (halves, quarters, and eighths). The duration of a

musical note is represented as a line segment, converted to a fraction (giving its fraction of a measure), and symbolized using standard musical notation. Students then use fraction addition to evaluate the accuracy of written musical parts.

Measures of Time, Part II applies the element of rate (tempo) and unit conversions to several authentic mathematics problems confronted by composers, conductors, and instrumentalists. For advanced classes or with some specified adjustments, Parts I and II can be combined into one class session.

Mathematics topics

Measurement, addition of fractions, symbolic representation. *Prerequisites:* Experience with fraction addition.

Music topics

Elements of standard rhythmic notation: measures, names and symbols for time values of notes, beats. *No prerequisites.*

Use with the primary curriculum

- In a prealgebra course to motivate the review of fractions
  Many students who know the algorithm for adding fractions may not really understand the concept. Measures of Time can provide an interesting way to see basic fraction relationships.

- Between units to demonstrate mathematics applications
  Measures of Time can be a fun and educational activity any time a special-interest application is appropriate.

Objectives

- To create a lasting and enhanced understanding of fractions
  Many students arrive in algebra classes without understanding fractions at a fundamental level. In Measures of Time students encounter and negotiate the essence of fraction relationships. This experience provides a drill that can correct the causes of fraction anxiety, poor performance, and misunderstanding.

- To inspire students to review fractions
  The context of Measures of Time is fun for students and immerses them in a review that might otherwise be dull.

- To expand awareness of the scope of mathematics
  Measures of Time shows students a strong connection between mathematics and music.

Student handouts

- Music and Mathematics: The Universal Languages (reading; one per student)

- Standard Music Rhythmic Notation (resource page; one per pair)

- Listening and Writing Fractions of Time (worksheet; one per student)

- Did These Composers Write Their Parts Correctly? (worksheet; one per student)

Materials

- CD tracks 11–22

- Blank scratch paper

- Overhead transparency of Listening and Writing Fractions of Time worksheet

- Blank timeline to demonstrate dictation

Instructional time

50–60 minutes

Instructional format

For the first half of this activity, you direct a discovery process for the whole class using the CD tracks. At each step of the activity script, key questions help you lead an inquiry that develops the system students will use to take dictation.

Do not pass out the worksheets until after students have created the timeline graphic (Steps 1 and 2). Once the timeline is established, hand out the worksheets and have students use the printed templates for dictation.

To combine Parts I and II of Measures of Time, you can omit Step 6, in which students sing a given rhythmic figure. You can also be more directive in the earlier steps, which establish the timeline template with students. To save more time, choose three or four dictation exercises rather than doing all six and let students do all the problems on Did These Composers Write Their Parts Correctly? as homework, or delete a few of the problems.

Student preparation

Have students read and discuss Music and Mathematics: The Universal Languages before the day of the activity to give them time to process the ideas. Refer to Step 1 of the activity script for suggestions about how to connect the ideas in the reading to the activity.

ACTIVITY SCRIPT

STEP 1   Introduce the activity and discuss the reading

Begin by reviewing these key ideas from the reading:

- Music and mathematics are universal languages that have their own symbolic systems.

- What is communicated by music is different from what is communicated by mathematics. (It can be fun, though abstract and murky, to probe the essence of this difference.)

(continued)

TEACHER NOTES

- The language of music (pitch and rhythm) is mathematical.
- This activity will explore one dimension of these languages and how they overlap—rhythm and rhythmic notation.
- This activity will require students to think mathematically about what they hear.

STEP 2 Create a graphic representation for sound in time

Students create a symbolic system for representing the time value of sound duration. The logic of the musical system is revealed to students through their own direct experience.

 Students in your class who have had musical training may find the beginning of this activity very basic. Urge them to follow along; the beginning of the activity does not last very long, and the dictation and mathematics problems will likely challenge them. Some of these students may not be high performers mathematically. This is an opportunity for them to be the experts in front of their peers.

Each student should have only a blank sheet of scratch paper on his or her desk. Play CD track 11, two cymbal crashes separated by about 20 seconds of silence.

Ask students:

What exists between the two cymbal crashes? [Time. Students may say that silence, reverberation, or nothing exists between the crashes. These are also good answers, but make sure they see that there is time between the cymbal crashes.]

How could we represent the time between cymbal crashes with a picture? [After some discussion, a student will suggest a line segment. Agree on this, and ask all students to draw a line segment to represent the passage of time between the two cymbal crashes.]

How could we represent the cymbal crashes? [Come to the conclusion of placing a hatch mark on each end of the line segment to serve as endpoints. Agree with students that this line segment will represent the total time that elapses between the cymbal crashes that they will hear.]

Inform students that musicians need a way to measure time very accurately.

Ask students:

How is time measured? [Seconds, minutes, hours, and so on.]

Acknowledge that these measures of time are used by musicians, but add that they also use a more versatile system that measures relative amounts of time. Play CD track 12 as an example of how musicians would create units to measure the time between cymbal crashes.

Ask students:

Are the bass drum beats evenly spaced between the cymbal crashes? [Yes.]

How many drum beats are there, including the beginning and the end? [7]

How many time intervals (smaller line segments) do the bass drum beats divide the total segment into? [6]

How might these intervals be easily represented? [Even marks on the line segment.]

Coach students to represent these smaller segments on their line by marking their endpoints with hatch marks. Conclude by pointing out that the bass drum marks units of time called *measures.*

STEP 3   Subdivide the measures

Play CD track 13.

Ask students:

What do you hear on the CD track now? [More subdivisions.]

Are the subdivisions equally spaced in time? [Yes.]

How could you indicate these subdivisions on your line segments? [With hatch marks. The end product should look like exercise 1 on Listening and Writing Fractions of Time.]

Conclude by indicating that each subdivision of the measure is called a *beat.*

The definition of the beat in relation to a piece of music varies from application to application. In music the beat is an agreed upon unwavering pulse against which time values of notes are measured. It can be assigned to any subdivision of the music based on what is more appropriate for the accurate interpretation of the music. One use of the term *rhythm* is to indicate a pattern of accents or sound in relation to the beat. The CD track including the bass drum is a simple rhythm.

TEACHER NOTES

STEP 4   Exercise 1: Rhythmic dictation

Discard the scratch paper and hand out the Listening and Writing Fractions of Time worksheet. Do not distribute the resource page yet. Explain that the next CD track is of a musical instrument playing with the rhythm. The students' task is to use a line segment to represent the amount of time that each note they hear is played. The line segments should be placed on top of the printed rhythm line for each example. The length of the line for each note should correspond to the number of beats that it lasts. The CD tracks use block chords with no change in pitch to avoid distracting from the element of rhythm. A snare drum part has also been added for musical interest. Track 14 is an example of this. Track 15 is the performance for exercise 1. Play the tracks and have students draw their line segments.

You will have to play the CD tracks several times. Some students may have a difficult time with this exercise. Suggest that they move their pencil along the line, unwavering, in sync with the beat (like the second hand of a clock sweeping around the dial), pushing down when and as long as a note is heard sounding. You can demonstrate this very effectively on an overhead transparency.

Use a copy of the worksheet on an overhead projector to show the solution, and have students correct their work if necessary.

STEP 5   Exercise 2: More dictation

Proceed exactly as in Step 4. Play the example and have students represent the notes with line segments on the rhythm line. Ignore the music staff below the line segment for now. It will be explained and used later.

STEP 6   Singing a line segment: Karaoke anyone?

Though this step is a good analogy to inverse operations in mathematics, it can be omitted. If you include it, you will find that it is fun and can help solidify the idea for students.

The process is now reversed: The note pattern is sung from a line segment representation. Here are two examples of one-note melodies represented as line segments:

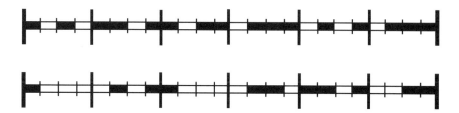

Place the first example on an overhead or on the board. Play the beat from track 14 and ask students to sing the rhythm using one note, as in the previous dictation exercises. Have them choose one syllable, such as "la" or "da." This could be done as a whole-class chorus or by a brave soloist. Ask a student with a musical instrument to play it using one note or any combination of notes. If this is a big hit and more students want a chance to perform solos, use the second example. You or your students can make up more rhythms to sing.

The musicians in your class will have a clear advantage, even though they probably have never read music from line segments before. This advantage can be an asset or a liability. Musicians can help lead discussions or provide hints and assistance to the other students. It will become a problem if the musicians in the class dominate the activity and intimidate the other students. Encourage them to help without dominating.

CD track 17 demonstrates a correct performance of the first example. Track 18 demonstrates the second example.

STEP 7   Time value names of notes as fractions

Turn students' attention to exercise 2 on their dictation worksheets. Explain that, although the line segments work well as visual pictures of the time values of notes, musicians use a system of symbols that is more useful to them and that corresponds very closely to the line segments we have used.

Ask students:
What fractional part of a measure is the first note of exercise 2? $\left[\frac{1}{2}\right]$
What fractional part of a measure are all of the notes in the example?

Have students write the fractions below the appropriate line segments for exercises 1 and 2.

Ask students:
If musicians wanted to name each type of note to indicate its length in time, what could they call the notes?

Coach them to arrive at the answers indicated on the resource page.

Musician students will immediately respond with the conventional terms quarter notes, half notes, and so on. However, it is good to ask them to withhold their answers, so that the vocabulary can be discovered directly by students. Many will respond with "one-fourth notes," "one-half notes," and so on. These students are essentially right. Show how such names were adapted to the conventional ones.

Pass out one resource page to each student pair.

STEP 8   Time value names for rests

Turn students' attention back to exercise 2. Review the fact that the unit of time is a measure.

Ask students:

If a note is not sounding, is time still passing? (You could have fun with this question.) [Yes.]

What should the total time of sound and silence in a measure add up to? [1]

Refer to the resource page and go over the symbols and terms for *rests* (periods of silence).

Have students place a fraction below each interval in exercises 1 and 2 in which a note is not sounding (no line segment) in the same way they indicated line segment fractions of the measure. Remind them that the sum of all the fractions for each measure must equal 1. These exercises are simple enough for students to easily check mentally.

STEP 9   Writing conventional rhythm notation

Now it is time to convert the line segments and rests to musical symbols. If necessary, explain that musical notation is written on a staff, as shown on the worksheet. The staff is similar to our rhythm line except that it has five lines. Tell students not to concern themselves with the five lines, as they are used to indicate pitch and that is not the subject of our study. Direct students to use the resource page to identify the appropriate symbols that represent notes and rests. On the staff for exercise 2 students should write these symbols directly underneath the rhythm line so that they line up with the appropriate line segments and rest intervals (as shown on the Answer key).

It is important to note some subtleties regarding this notation. As stated on the resource page, a tie indicates that the time values of the tied notes are added, with the notes sounding as one note lasting the length of the sum of the tied notes. This gives rise to multiple ways to notate the same sound. For example, a half note sounds identical to two tied quarter notes. Having these very different looking symbols for the exact same sound can cause confusion. As students notate the exercises, the class will produce a variety of different technically correct answers. This is fine. In music practice, preferred conventions serve the needs of visual organization of the rhythms in relation to the beat for clarity. Considering these conventions, while interesting and informative, is unnecessary for this activity.

TRACKS
19–22

STEP 10   Exercises 3–6: More dictation

For the remaining exercises, follow the same process as in Steps 3 and 4, but ask students to apply music notation on the staffs after they have drawn the line segments. Use CD track 19 with exercise 3; CD track 20 with exercise 4; CD track 21 with exercise 5; and CD track 22 with exercise 6.

The rhythm behind the notes on the CD tracks now has further subdivisions of eighth notes played in the background. These are included to prepare students for exercises 5 and 6 and to provide more musical interest.

You may neither need nor want to do all the exercises. The exercises increase in difficulty so that you can adapt the activity to the interest and skill level of your class.

Musicians count time according to the beat number and repeat their count for each measure. Since a quarter note is a beat for our exercises, counting beats means counting quarter notes. The counting is done silently as "one, two, three, four, one, two, three, four, . . ." as the beats pass, four counts per measure in this case. Musicians can assign a different number of notes to a measure; this is discussed in the next step.

In exercises 5 and 6, the rhythmic figures are broken into eighth notes. You will need to pause and discuss this with your students. Have them focus on the background eighth-note pulse in the rhythm of the CD track to assist them in their dictation. To count eighth notes, they should place the word *and* between number counts.

STEP 11   Did These Composers Write Their Parts Correctly?

After you get them started, students can answer the remaining questions in pairs or alone as a homework assignment.

Direct hints are purposely left off the worksheet to force students to make appropriate connections and draw on the knowledge gained in class. If some students have difficulty making connections, lead them to the process of converting each note to its fractional equivalent and writing the fraction below the note. Add the fractions to verify that the sum is 1. If the sum of note lengths in a measure is not 1, there's an error.

Before students can complete exercises 5 and 6, you will need to present the concept of time signatures. An explanation of the concept is provided in the following discussion point. At a minimum, students need to understand that a three-four time signature assigns three quarter notes to each measure, requiring the sum of all notes and rests to equal $\frac{3}{4}$ instead of 1 as in the previous exercises. A five-four signature assigns five quarter notes per measure, requiring the sum of the notes and rests to equal $\frac{5}{4}$ for the part to be correct.

TEACHER NOTES

The system used thus far is very logical—quarter notes are a fourth of the measure, half notes are a half, and so on. Musical convention uses a much greater diversity. Any number of notes of any type can be assigned to a measure as the basis to organize rhythm. Examples include three quarter notes per measure, two quarter notes per measure, six quarter notes per measure, and three half notes per measure. This assignment is called a *time signature* and is indicated at the beginning of a piece of music. It becomes the rule that determines what type of note will represent the beat and how many beats will be assigned to each measure. In time signatures other than four-four, the pure mathematical logic of a quarter note being a quarter of something no longer applies.

A time signature is written on the staff like a fraction without the division line. The top number indicates the number of beats per measure, and the bottom number indicates what type of note will be used to represent the beat. A composer's decision as to what time signature to use involves many considerations, including what will be most easily read by the conductor and musicians and the kind of rhythmic feel that is desired. It is common for a time signature to change within a piece·of music.

FOLLOW-UP ACTIVITIES

Measures of Time, Part II

Measures of Time, Part II uses the knowledge of rhythmic notation from Part I to pursue a variety of authentic problems faced by instrumentalists, composers, and conductors.

Textbook assignments

It is valuable to follow Measures of Time with skill practice in manipulating fractions and with applications of fractions in different contexts.

Writing prompts

- How is communication in mathematics and music similar? How is it different?

- Did the activity change how you see mathematics and music? How?

- Describe what you think the difference is between the kind of knowledge that mathematics communicates and the kind of knowledge that music communicates.

- Have you ever been able to use mathematics or music to communicate with someone who spoke a different language?

Research projects

- Students can research the mathematics of more advanced aspects of rhythm, such as triplets or septuplets and where they occur in musical pieces. Perhaps they can collaborate with musicians in the class to present what they have learned.

- Students can research how musical notation and time signatures evolved.

ANSWERS

Listening and Writing Fractions of Time

1.

2.

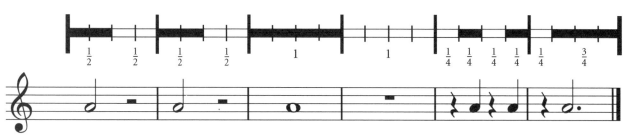

3.

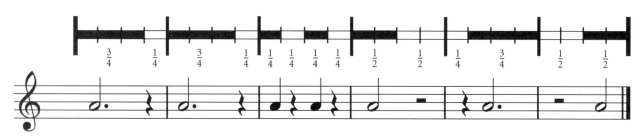

4.

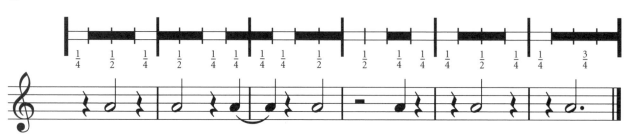

5.

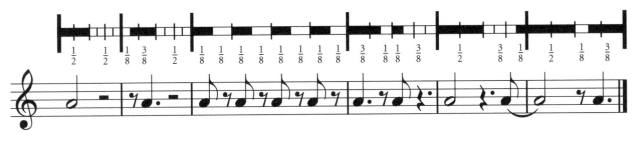

6.

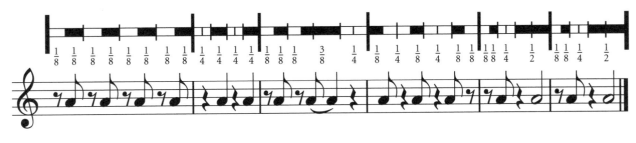

Did These Composers Write Their Parts Correctly?

1. $\frac{1}{2} + \frac{1}{8} + \frac{1}{4} + \frac{1}{8} = 1$; correct

2. $\frac{1}{4} + \frac{1}{8} + \frac{1}{8} + \left(\frac{1}{4} + \frac{1}{8}\right) + \frac{1}{4} = \frac{9}{8}$; $\frac{9}{8} > 1$; incorrect

3. $\frac{1}{8} + \left(\frac{1}{4} + \frac{1}{8}\right) + \left(\frac{1}{2} + \frac{1}{4}\right) = \frac{5}{4}$; $\frac{5}{4} > 1$; incorrect

4. $\frac{1}{8} + \frac{1}{8} + \left(\frac{1}{4} + \frac{1}{8}\right) + \frac{1}{4} + \frac{1}{8} = 1$; correct

5. $\left(\frac{1}{4} + \frac{1}{8}\right) + \left(\frac{1}{8} + \frac{1}{16}\right) + \left(\frac{1}{8} + \frac{1}{16}\right) = \frac{3}{4}$; correct

6. $\frac{1}{2} + \left(\frac{1}{4} + \frac{1}{8}\right) + \frac{1}{8} + \frac{1}{8} = \frac{9}{8}$; $\frac{9}{8} < \frac{5}{4}$; incorrect

TEACHER NOTES

MUSIC AND MATHEMATICS: THE UNIVERSAL LANGUAGES

Did you ever think of mathematics as being a language? What is a language? Language is defined as a systematic form of communication. Mathematics uses distinct symbols to communicate ideas about numbers, shapes, actions, relationships, and abstract concepts. It uses symbols to communicate just as verbal languages use words. Equations can actually be thought of as mathematical sentences with nouns and verbs. As it turns out, mathematics is a systematic form of communication—a language—shared by people of many nationalities.

What about music? Is music also a systematic form of communication? Composers often write music to communicate ideas or emotions too abstract to describe in words. We have been "spoken to" by music when we feel different after listening to it. You might say that notes, melodies, and rhythms make up the language that musicians use to communicate. Music is in fact a language, just like verbal language or mathematics.

Although music and mathematics communicate different types of messages, they are both understood by people from many cultures. The ways in which notes, melodies, and rhythms are used in music vary greatly depending upon musical style, but the notation system is the same for many nationalities. The same is true for mathematics. Both music and mathematics bridge cultural barriers and enable people who speak different verbal languages to have musical and mathematical "conversations" in a common language.

Music and mathematics are also similar in some of the ways their symbol systems relate to each other technically. Musical notation is the system of written symbols that communicates notes and rhythms and how to play them. In the same way that these written symbols are not what actually moves us about a piece of music—we have to hear the music and feel it—mathematical symbols are not the mathematical concepts themselves, but written representations of those concepts. The written symbols for music look very different from mathematical symbols, but some of them are in fact similar in what they represent. Some musical symbols actually represent mathematical quantities—quantities of time and distances between notes. Musicians and composers work with and calculate the relationships of these quantities when they write and perform music. The musical staff is actually designed like a graph that you use in math class. This makes musicians a lot like mathematicians!

It's fascinating to consider that while music and mathematics communicate such different messages, they are similar in the way they are constructed as languages and in how they are used throughout the world as systematic forms of communication. In the activity Measures of Time you will explore one piece of the musical language—how musical notation mathematically measures and represents the length of time that you hear musical notes.

STANDARD MUSIC RHYTHMIC NOTATION

Name	Fraction of a measure (in $\frac{4}{4}$ time)	Symbols Sound (notes)	Silence (rests)
Whole note	1	𝅝	▬
Half note	$\frac{1}{2}$	𝅗𝅥	▬
Quarter note	$\frac{1}{4}$	♩	𝄽
Eighth note	$\frac{1}{8}$	♪	𝄾

Special operations on note length

Ties—A tie between two notes combines the two notes into one note. The length of the resulting note is the sum of the lengths of the two tied notes. Example:

$$\frac{1}{2} \quad + \quad \frac{1}{4} \quad = \quad \frac{3}{4} \qquad\qquad \frac{1}{4} + \frac{1}{8} = \frac{3}{8}$$

Dotted notes—A dot after a note increases its length by half its value. Example:

$$\frac{1}{2} \quad + \quad \frac{1}{4} \quad = \quad \frac{3}{4} \qquad\qquad \frac{1}{4} + \frac{1}{8} = \frac{3}{8}$$

LISTENING AND WRITING FRACTIONS OF TIME

Under the direction of your teacher, use the music examples and your resource page to complete the exercises on this page with your partner.

1.

2.

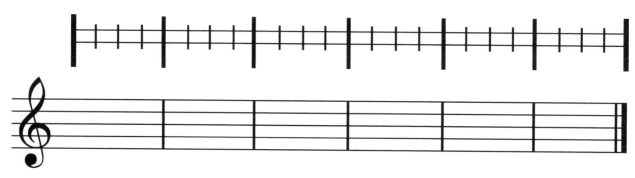

3.

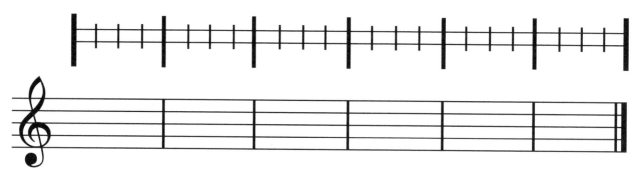

4.

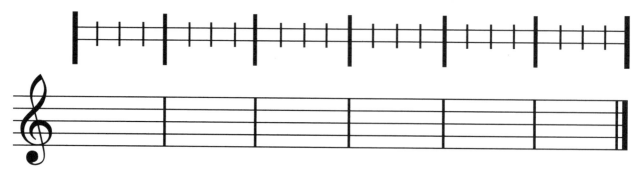

Listening and Writing Fractions of Time (continued)

5.

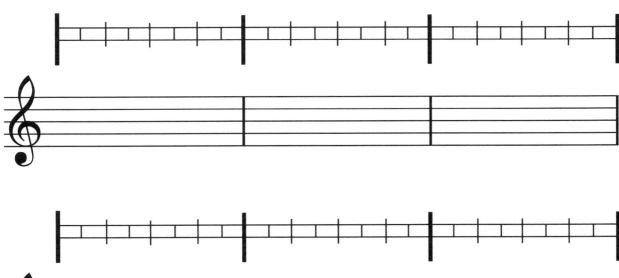

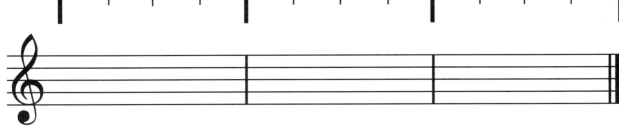

6.

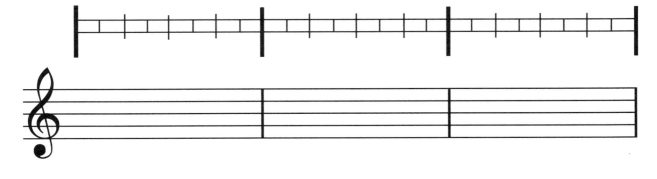

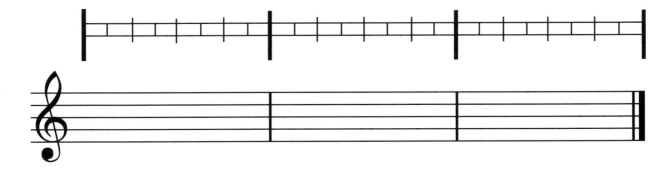

DID THESE COMPOSERS WRITE THEIR PARTS CORRECTLY?

For each measure below, determine whether the part is written correctly and justify your answer with appropriate calculations. Use your knowledge about measures, notes, and fractions from today's activity.

1.

2.

3.

4.

5.

6.

3

Measures of Time, Part II

Tempo and Rate Problems Facing Musicians

Measures of Time, Part II applies the concepts developed in Part I to a series of authentic problems. The instructional format is relatively simple. Students begin with identifying the tempo of the music on the CD track. Then they think and talk about how that tempo could be measured. Students develop the notion of tempo as the rate of the beat and work either in pairs or individually to solve some problems. They use the Standard Music Rhythmic Notation resource page from Measures of Time, Part 1 to compare a tempo to such things as beats per second or the speed at which a musician can play a sixteenth note. The problems they solve involve situations encountered by composers, conductors, and instrumentalists.

TEACHER NOTES

Mathematics topics

Rates, unit conversions, dimensional analysis, operations on fractions. *Prerequisites:* Experience interpreting word problems, fraction multiplication and division, basic experience with unit analysis.

Music topics

Tempo, rhythmic notations. *Prerequisites:* An understanding of concepts from Measures of Time, Part I (fundamental concepts of musical rhythm notation, note values, and measures).

Use with the primary curriculum

- As enrichment while studying unit conversions or rate problems Measures of Time, Part II provides a rich context for practicing and expanding the use of dimensional analysis as a problem-solving tool. It applies units of measurement and unit fractions that will be new to most students. For this reason it is not recommended as an introduction to operations on units.

Objectives

- To provide practice with problem-solving strategies using dimensional analysis
- To provide real-world applications of the skills and concepts developed in Measures of Time, Part I

Student handouts

- Tempo Problems of Musicians (worksheet; one per student)
- More Tempo Problems (worksheet; one per student)
- Standard Music Rhythmic Notation (resource page for Part I on page 25; one per group)

Materials

- CD track 74

Instructional time

20–50 minutes

Instructional format

The activity begins with the entire class determining the tempo of the music on a CD track. After the tempo concept is established, students can work in pairs, in groups, or individually.

Student preparation

Measures of Time, Part I is sufficient preparation. If you have not done Part I, you will need to directly teach the rhythmic concepts developed in Part I. These concepts are summarized on the resource page Standard Music Rhythmic Notation. You should also give students the reading from Part I, Music and Mathematics: The Universal Languages, and discuss the ideas in it.

If you want students to use dimensional analysis in solving the problems, it can be valuable to precede this activity with some lessons that apply unit conversion and unit analysis. Many students are uncomfortable multiplying and dividing by unit fractions, such as 60 seconds per minute. Some recent experience with other unit fractions, such as 12 inches per foot, will enhance the effectiveness of this activity.

ACTIVITY SCRIPT

STEP 1   Establish the concept of tempo

TRACK 74

Begin by asking students if they know what tempo is. [*Tempo* is the speed at which a piece of music is played.] Then ask students for examples of tempos and discuss the effects of tempo on the listener.

> *Ask students:*
> How do you think the tempo of a piece of music could be measured
>   and indicated? [Answers will vary. Suggest the following question.]
> What kind of unit is generally used to measure heart rate? [Beats per
>   minute.]
> Does it make sense to use these units to measure the rate of a piece
>   of music? [Yes.]

Conclude that the tempo of a piece of music is indicated by the number of beats per minute.

As you play CD track 74, students will use a stopwatch or the second hand on a wristwatch to count the number of beats in 1 minute of the rhythm they hear. You will, of course, have to agree on what to count as the *beat*.

Ask students for a quicker way to find the beats per minute than counting for a whole minute. Counting for only 15 seconds and multiplying by 4 or counting for 30 seconds and multiplying by 2 will be technically correct, but which method might result in greater accuracy? Counting beats over smaller intervals will compound error more dramatically when students calculate the beats per minute. Agree on a happy medium.

If you are combining Part I and Part II, you can skip Step 1, but be sure to define *tempo* and explain *beats per minute*.

STEP 2　Tempo problems

Direct students' attention to the worksheet Tempo Problems of Musicians. Read through the tempo facts at the beginning and let students know that they will be multiplying or dividing by unit fractions such as $\frac{1 \text{ min}}{60 \text{ sec}}$ and $\frac{4 \text{ quarter notes}}{1 \text{ whole note}}$. Then let them work on their own. Depending on the background of your students, you may need to provide only hints and assistance. Only after you get students on track are they likely to be able to complete the problems successfully. The problem on the second worksheet, More Tempo Problems, can be assigned as homework.

Work these problems yourself before viewing the solutions provided. There are many different ways that they can be approached, and your insights into the difficulties your students may have will be enhanced by your doing the problems yourself. Many students will not approach the problems using units in a systematic way but will manage to solve the problems by their own unorthodox methods, often hit and miss. Letting students share their work and compare strategies can be valuable.

FOLLOW-UP ACTIVITIES

Textbook assignments

The work begun in Measures of Time, Part II is solidified by students working similar problems that apply unit conversions or rates in different contexts.

Projects

- Obtain music scores with tempo markings, and compute the time for various sections of the music. Invent a series of problems to present to the class.

- Interview musical directors or conductors to find out what kinds of mathematical problems they actually encounter and how many solutions they estimate by using mental mathematics.

ANSWERS

Tempo Problems of Musicians

1. Faster

2. $100\,\dfrac{\text{♩}}{\text{min}} \cdot \dfrac{1\text{ min}}{60\text{ sec}} = 1.7\,\dfrac{\text{♩}}{\text{sec}}$

   Or think: You want $\frac{\text{♩}}{\text{sec}}$, so you need ♩ in the numerator and seconds in the denominator. Thus $100\,\frac{\text{♩}}{\text{min}}$ times what unit fraction will give $\frac{\text{♩}}{\text{sec}}$? Minutes will have to be in the numerator, so the unit fraction is $\frac{1\text{ min}}{60\text{ sec}}$.

3. $\dfrac{100\,\frac{\text{♩}}{\text{min}}}{4\,\frac{\text{♩}}{\text{o}}} = 25\,\dfrac{\text{o}}{\text{min}}$

4. $\dfrac{60\,\frac{\text{sec}}{\text{min}}}{25\,\frac{\text{o}}{\text{min}}} = 2.4\,\dfrac{\text{sec}}{\text{o}}$

5. $\dfrac{2.4\,\frac{\text{sec}}{\text{o}}}{8\,\frac{\text{♪}}{\text{o}}} = 0.3\,\dfrac{\text{sec}}{\text{♪}}$

6. ♩ + ♩. = 7 eighth notes, and an eighth note lasts 0.3 sec; therefore ♩ + ♩. lasts 0.3(7) = 2.1 sec.

More Tempo Problems

1. $\dfrac{50\,\text{♩}}{\text{min}}\left(4\,\dfrac{\text{♪}}{\text{♩}}\right) = 200\,\dfrac{\text{♪}}{\text{min}}$

   $\dfrac{200\,\frac{\text{♪}}{\text{min}}}{60\,\frac{\text{sec}}{\text{min}}} = 3.3\,\dfrac{\text{♪}}{\text{sec}}$

2. $\dfrac{4\,\frac{\text{♩}}{\text{o}}}{3\,\frac{\text{sec}}{\text{o}}} = \dfrac{4}{3}\,\dfrac{\text{♩}}{\text{sec}}$

   $\left(\dfrac{4}{3}\,\dfrac{\text{♩}}{\text{sec}}\right)\left(60\,\dfrac{\text{sec}}{\text{min}}\right) = 80\,\dfrac{\text{♩}}{\text{min}}$

   or ♩ = 80

   In one step: $\left(\dfrac{60\text{ sec}}{1\text{ min}}\right)\left(\dfrac{\text{o}}{3\text{ sec}}\right)\left(4\,\dfrac{\text{♩}}{\text{o}}\right) = 80\,\dfrac{\text{♩}}{\text{min}}$

3. $\left(190\,\dfrac{\text{♩}}{\min}\right)\left(\dfrac{2\text{ notes played}}{\text{♩}}\right) = 380\,\dfrac{\text{notes played}}{\min}$

$$\dfrac{380\,\dfrac{\text{notes}}{\min}}{60\,\dfrac{\sec}{\min}} = 6\,\dfrac{\text{notes}}{\sec}$$

4. When sixteenth notes are being played, four notes are played for every quarter-note beat, so:

   Band's maximum rate: $= \left(140\,\dfrac{\text{♩}}{\min}\right)\left(4\,\dfrac{\text{notes played}}{\text{♩}}\right) = 560\,\dfrac{\text{notes played}}{\min}$

   When eighth notes are being played, four notes are played for every half-note beat, so:

   Song tempo: $\left(145\,\dfrac{\text{♩}}{\min}\right)\left(4\,\dfrac{\text{notes played}}{\text{♩}}\right) = 580\,\dfrac{\text{notes played}}{\min}$

   The musicians can't play fast enough.

5. Musicians' maximum rate:

   $$\left(560\,\dfrac{\text{notes played}}{\min}\right)\left(\dfrac{\text{♩}}{4\text{ notes played}}\right) = 140\,\dfrac{\text{♩}}{\min}\ \text{ or }\ \text{♩} = 140$$

TEMPO PROBLEMS OF MUSICIANS

Tempo facts

- The *tempo* of a song is how fast the beats occur. In music, this is expressed as beats per minute.

- Tempo notation includes the beats per minute and the type of note that represents a beat. For example, a tempo of 120 beats per minute with a quarter note used to represent the beat is written simply as ♩ = 120. Notice that the phrase *beats per minute* is omitted.

Tempo problems

A producer is reviewing a rap song that has a tempo of 100 beats per minute. In this song, each beat is represented by a quarter note, so the tempo is 100 quarter notes per minute. This is indicated in music as ♩ = 100. The six problems below use this tempo. Some answers may be fractions of notes or fractions of seconds. Express these answers as decimals rounded to the nearest tenth.

1. To get a feel for the designated tempo of a song, musicians find it useful to compare the rate of the beat to how fast seconds go by. Are the beats of the rap song faster or slower than passing seconds?

2. How many quarter notes per second is this tempo?

3. How many whole notes per minute is this tempo?

4. How many seconds would a whole note last?

5. How many seconds would an eighth note last?

6. How many seconds would a half note tied to a dotted quarter note last?

MORE TEMPO PROBLEMS

1. A song has a tempo in which a half note equals 50 beats per minute. How many eighth notes would be played per second at this tempo?

2. A composer is working with a melody he likes. He makes an artistic decision that the whole notes should last for 3 seconds. They just sound good to him that way. What tempo marking would he have to use, in quarter notes per minute, so that the whole notes will be 3 seconds long when the piece is played?

3. A flute player explains to the conductor that the fastest she can play eighth notes on her instrument is at a tempo of 190 beats per minute (where a quarter note gets the beat). How many eighth notes per second can this musician play? Round your answer to the nearest note. (*Hint:* Consider how many notes per quarter-note beat the flute player can play.)

4. A conductor has found that her band generally cannot play sixteenth notes if the tempo is any faster than 140 quarter-note beats per minute. (Sixteenth notes are one sixteenth of a measure, or twice as fast as eighth notes.) She is considering a piece of music that does not contain any sixteenth notes but does contain many eighth notes. The tempo is indicated as ♩ = 145. Can the musicians play the eighth notes at the tempo indicated? Show work to justify your answer. (*Hint:* Consider notes the band can play per minute and the number of eighth notes that must be played per minute in the piece of music.)

5. At what tempo could the band in problem 4 play the piece (in half notes per minute)?

4

The Multiples of Drummers

The Mathematics of Polyrhythms

This fun activity challenges students' motor coordination and auditory perception and helps them discover a rule for finding the least common multiple (LCM) of a set of numbers. Student groups clap various rhythmic cycles simultaneously and represent them symbolically on a polyrhythm chart. Polyrhythms created from the superimposed rhythms are performed, listened to at various tempos, and displayed on the chart to show the relationships between factors and multiples that lead to a general rule for finding the LCM of any set of numbers. After the performances, a variety of applied music problems involving polyrhythms provide multicultural perspectives and present some powerful extension problems and explorations.

The performance nature of the activity provides a different access point for the review of multiples and the LCM. The entire class is engaged in an interactive performance, physically experiencing shifting resonance points between different clapping patterns. Regardless of students' mathematical backgrounds, The Multiples of Drummers provides a fresh view of multiples and a unique experience.

Mathematics topics

Counting, multiples, least common multiples, factoring, exponents, ratio, patterns, problem solving. *Prerequisites:* Knowledge of the concepts of prime factors and multiples.

Music topics

Beat, rhythm, polyrhythms, phrasing, tempo, accents, performance. *No prerequisites.*

Use with the primary curriculum

- To introduce the LCM for prealgebra
  This activity can provide a way for students to develop the rule for the LCM in a prealgebra course.

- To review the LCM for algebra I
  Use this activity before lessons on the LCM of algebraic expressions or on adding and subtracting algebraic fractions. Solid understanding of and skill with LCMs are fundamental to efficient manipulation of algebraic fractions, and a weakness in this area can undermine a student's success well into calculus courses.

- As a special project
  The Multiples of Drummers can be used to set the context for the extensions as special projects or problems of the week.

- For review, enrichment, or assessment
  The Multiples of Drummers can provide a powerful assessment or enrichment tool after the concepts of multiples and the LCM have been taught using another curriculum.

Objectives

- To enhance understanding of multiples and factors
  Practice with using the relationships of multiples in a unique context can improve understanding for all students.

- To strengthen retention of concepts
  When students discover mathematical rules and concepts through observation and active experimentation, they remember them longer.

- To provide access to mathematical concepts for students of various learning styles
  The combination of kinesthetic, visual, and aural representations of quantitative relationships can provide access for learners with strengths in these areas.

- To foster multicultural insights
  Mathematics is used as a tool to help students gain new appreciation and understanding of the art of different cultures.

Student handouts

- The World Is a Drummer (reading; one per student)
- Help and Information (resource page; one per pair)
- Polyrhythm Chart and Problems (worksheet; one per student)
- Polyrhythms in Music (worksheet; one per student)

Materials

- CD tracks 23–28
- Metronome (optional)

Instructional time

30–60 minutes

Instructional format

This activity can give you an opportunity to fulfill a possible fantasy—to conduct a 20-piece music ensemble. Conductor will be your primary role. To begin, divide the class into three large groups. These groups will perform different rhythms by clapping to a metronome beat. The entire class works together to perform and analyze an assortment of rhythms. Use an overhead transparency of the polyrhythm chart, and mark rhythms during the whole-class analysis of the performances.

Following the whole-class performances, student pairs use the resource page to work through a set of problems guided by your coaching. Periodically the entire class comes together for debriefing to ensure that all students are reaching accurate conclusions. The final set of problems, Polyrhythms in Music, can be completed by students in pairs, with little assistance from you—or it can be assigned as homework.

Remember to review the content of the resource page so that you can adjust the amount of direct instruction you deliver to match your teaching style, the extent to which you intend to create a constructivist experience for your students, and the level of independence your students can handle.

Student preparation

As is the case with all the activities in the book, it is important to set the context for, and warm students up to, the type of activity they are going to be doing. Have them read The World Is a Drummer either the night before or in class prior to the lesson. If your class has been doing several other music activities and you are short on time, you could skip this reading.

ACTIVITY SCRIPT

STEP 1　Introduction and background information

Review and discuss the reading and give students an overview of the activity. Explain that they will clap rhythms and chart their performances to discover the mathematics of rhythm. Don't forget to emphasize that they should have fun in the process!

At this point introduce the following basic musical concepts and terms:

Musicians perform music in reference to a steady pulse called the *beat*. In musical practice the beat is played in some form by a musician, directed by a conductor, or just imagined in the musician's mind. In this activity it is represented by a steady metronome pulse. The speed of the beat is called the *tempo*. Patterns of accents on the beats create *rhythms*. Combining two or more different rhythms creates *polyrhythms*. We will be concerned with finding the phrase length of rhythms and polyrhythms, which is measured by the number of beats it takes for the rhythm to repeat its cycle.

 Don't get bogged down with trying to ensure complete understanding of the background information. The meaning of these ideas will be clarified after the first performance.

The tempo of musical pieces is indicated in beats per minute to conductors and musicians. The performance tempo for this activity is 100 beats per minute.

STEP 2　Create groups and assign roles

Here are two options for student groupings and roles depending on the mathematical background of your students.

Option 1—prealgebra and algebra 1: Divide the class into three equal groups. Two of the groups will be clappers performing two different rhythm cycles simultaneously, and the third group will be observers. The observers will count beats and look for the beat numbers on which the two clapping groups clap together. The clapping groups will be referred to as group A and group B. As each new polyrhythm is performed, it is important to have students shift roles so that all students have equal time clapping. This can be time consuming, however, since having new groups learn parts for each performance can become tedious.

Option 2—algebra 1, algebra 2, and above: Divide the class in half to make two even clapping groups and eliminate the observer group. For more advanced students, the observer group is unnecessary—it will be obvious to most of these students even before the performance that the resonance point is the LCM of the two primary rhythms.

Be creative by suggesting that students use body movements or spoken words instead of clapping. To make the activity a visual and kinesthetic experience, have students in clapping groups raise both hands suddenly at a point where they otherwise would have clapped. Students could also shout a word at the same instant, or they could skip the hand movements altogether and use only words. These variations can be great fun for the class and very valuable for kinesthetic and visual learners.

STEP 3   Perform the first polyrhythm, 2:3

The process described below will be repeated throughout the activity with different rhythms.

Explain that all the clapping will be done in reference to the metronome beat. Play the metronome on CD track 23 or use a metronome set at 100 beats per minute.

Group A will clap the first rhythm listed on the chart, 1:2. This ratio indicates one clap every two beats, so the group A clappers will be clapping on beats 0 (go), 2, 4, 6, and so on. Play the metronome beat and have the students practice this until they are clapping together; then stop and instruct group B.

Group B will clap the second rhythm listed on the chart, 1:3. This ratio indicates one clap every three beats, so group B clappers will clap on beats 0 (go), 3, 6, 9, and so on. As with the previous group, play the metronome and have students clap their rhythm until they are together and confident.

Groups A and B now clap their rhythms simultaneously. You will have to conduct this. Cuing students to start together can be a little tricky. The best way to do it is to let the metronome run for a few beats and then give a "one, two, go" instruction with students beginning their clapping on the word *go*. Giving this count on every other beat is the easiest to hear and comprehend for all involved. It may take a few attempts to coordinate the two groups.

If you are using observers, have them take note of what beat numbers groups A and B clap on together. Or have the clapping groups identify the beat number of the common claps. The distance between the simultaneous claps is the phrase length.

The rhythm made of two separate rhythms, 1:2 and 1:3, is called a *polyrhythm*. The resulting polyrhythm in this case is 2:3.

The polyrhythm 2:3 is a very common polyrhythm found in many types of music, from rock to classical to jazz. Playing a three-beat phrase in the space of a two-beat phrase is called a *triplet* in musical terminology. In many African cultures, it is believed that 3 is a male number and that multiples of 2 are female numbers. It is also believed that both male and female elements need to be present to reach perfection in any endeavor, leading to the presence of 2:3 polyrhythms in much of African music.

Students now fill in the polyrhythm chart for the rhythms they have performed. Have them place marks on the appropriate beat lines to signify claps. Using a different symbol for each polyrhythm can help students visually differentiate the polyrhythms. Confirm with the class that the phrase length for a 2:3 polyrhythm is 6. Consult the observers for verification, and direct all students to fill in the phrase length in Table 1.

Play the recorded demonstration on CD track 24. This demonstrates the 2:3 polyrhythm played perfectly and also at faster tempos. Notice that as the tempo increases, the rhythm sounds less awkward and almost familiar. You may want to bring in other obvious musical examples.

TRACK 24

Students will determine the phrase length in various ways: from their mathematical concept of common multiples, from the observer group, and from the polyrhythm chart. If discussion reveals varying methods and perceptions, it can be valuable to have students share these with the class. Even for students who see the phrase length as obvious from the outset, this simple problem is a useful starting point because the problems get more difficult.

STEP 4   Perform the next polyrhythm, 3:4

Follow the process outlined in Step 3 to perform the 3:4 polyrhythm.

As previously suggested, you may choose to rotate the roles of the groups. If not, it is efficient to let the group clapping the 1:3 rhythm continue to do so and then have the other group switch from clapping 1:2 to clapping 1:4.

After students perform the 3:4 polyrhythm, play CD track 25 to demonstrate the sound of the 3:4 polyrhythm played perfectly. Do students think it sounds more musical as the tempo increases?

TRACK 25

A good example of music with this polyrhythm is "Fantasy Impromptu" by Chopin, in which the right hand plays four beats to the left hand's three.

Have students fill out the polyrhythm chart for these rhythms, along with Table 1.

The polyrhythm chart closely models how pitch frequencies interact. Musical notes are created by pulses, like clapping, happening at very fast frequencies. Different pitches with frequencies that have common factors sound more harmonious to our ears than those with no common factors. An interesting parallel to this phenomenon exists with polyrhythms. Rhythms that fit well with each other seem to have phrase lengths with common factors. This idea is discussed in more detail in the extensions at the end of this activity script. The numerical relationships between pitch frequencies is the subject of Scaling the Scale, Part I.

STEP 5   Complete the remaining polyrhythm performances and polyrhythm problems

You can choose to perform as many of the remaining polyrhythm possibilities as you like. CD track 26 has polyrhythm 4:5, and CD track 27 has polyrhythm 2:3:5.

Feel free to experiment with some other combinations. For the purposes of the activity, the entire polyrhythm chart should be filled out along with Table 1 before you proceed to the polyrhythm problems. I recommend that students perform at least 4:5 and 2:3:5. For 2:3:5, you will need three clapping groups. Note that Table 1 contains the polyrhythm 3:5, which is not on the CD. Students will need to think about the mathematics rather than just recording what they observe.

When the chart is complete, pose the question of what it would sound like if all the rhythms were played simultaneously: polyrhythm 2:3:4:5:6:7. Intuition might suggest that this would sound like chaos, considering that there are few common factors and that the phrase length would be very long. If you really feel ambitious, you might try having students perform it. At least play CD track 28 to demonstrate 2:3:4:5:6:7. Do students think it sounds chaotic?

Some students will want to work out the phrase length of 2:3:4:5:6:7. This is fine, but put off presenting solutions or giving hints until problem 3 of Polyrhythms in Music is encountered. (The phrase length is 420 beats.)

Dance in many African cultures incorporates complex polyrhythms. A good dancer will define up to four different rhythms at once, one with each limb! On first glance, these dances may appear to consist of random movements, but careful observation reveals a high level of sophistication and order.

After some discussion with the class, play the CD track. It is surprisingly musical, and one could actually dance to it. It has the quality of African rhythms that use many polyrhythms simultaneously.

Pause to have students analyze why the sound of all these rhythms together has a sense of cohesion and why it sounds organized. For one thing, many of the rhythms within it have short phrase lengths, and the listener is drawn to those. The short lengths form a backbone for the structure. The phrases that take many beats to repeat are heard almost as colorful accents and do not distract from the organization of the shorter phrases.

STEP 6   Analysis: Polyrhythm problems

At this point, the activity mode shifts from performance to analysis. It can be effective to maintain the whole-class engagement and work through these problems in a discussion format as a class. Feel this out. Depending on the personality and background of your class, you may decide to work through these problems with students in pairs. The questions form a directed discovery sequence that ends with problem 7, which defines the general rule for determining the least common multiple of any set of numbers or algebraic expressions.

It is not expected that students will reach an exact rule on their own purely from the information obtained in the activity. The extent to which this rule is student-generated will vary from class to class and within a class. This sequence of problems needs to be supplemented with your gentle guidance and leading questions. Remind students to use the resource page for more help. Bring the class together after problems 5, 6, and 7 to share conclusions. Before moving on, be sure all students have an accurately articulated version of the general rule for finding the phrase length of any polyrhythm, this being the rule for finding the LCM for a set of numbers.

To help students discover the rule for finding the LCM, bring the class together and lead them through some more examples of finding the LCM for numbers that contain common factors. As students observe the answers for a set of problems, they can find a pattern to create the rule. First establish what the LCM means. Then, with the class, express each number and the LCM in prime factorization form using exponents, and look for patterns. Present as many examples of this as necessary. For example:

The LCM of {10, 15, 12} is 60.

Students factor these numbers into primes:

$$10 = 2(5) \qquad 15 = 3(5) \qquad 12 = 2^2(3) \qquad LCM = 2^2(3)(5)$$

Compare the factors of the LCM with the factors in the set of numbers. Notice that the LCM is the product of the greatest power of each prime factor in the set of numbers.

STEP 7   Polyrhythms in Music

Students can work on their own to complete the applications in Polyrhythms in Music. Problem 4 presents a complex mathematical relationship that cannot be worked out using simple multiples. Suggest that students do the problem graphically, and ask them what is different about it. Notice that the answer, 22, can be determined mathematically by $2(5) + 3(4)$. While this suggests a general formula, it works only for particular sets of numbers. This problem is included to introduce another level of complexity and to provide a launching point for an extension problem.

FOLLOW-UP ACTIVITIES

Textbook assignments

It is important to follow The Multiples of Drummers with some skill practice that requires finding LCMs either numerically or algebraically. Adding and subtracting fractions or pure LCM exercises are given meaningful context by this activity, and practice will help solidify the pure mathematical skill and understanding offered by the activity.

Writing prompts

- What did you learn in today's activity?
- What music that you are familiar with uses some of the polyrhythms you clapped today in class?
- Which polyrhythms sounded the most familiar? The least familiar?
- Which polyrhythms were the easiest to perform? Explain why in mathematical terms.
- Which polyrhythms were the most difficult to perform? Explain why in mathematical terms.
- Did today's activity help you understand more about multiples? Why? How?

Extensions

- Connection to pitch and frequency
  Explore and research how the polyrhythm chart models the interaction of different pitch frequencies. When two different pitches sound simultaneously, a perceptible pulse (a dull throbbing) is heard, called the *beat frequency.* This frequency is the difference of the frequencies of the two pitches. It can be interesting to compare this frequency relationship to the

relationships in polyrhythms. Our activity considers the beat as a measure of time and observes repetitions of a rhythmic cycle (phrase) in relation to it. The frequency of repetition of a phrase can be expressed as the number of cycles (phrase repetitions) that occur per unit of time, in this case, cycles per beat. Thus the rhythms have frequencies of $\frac{1}{2}$, $\frac{1}{3}$, and so on. This prompts the following question: Is the relationship of the phrase length of a polyrhythm to the lengths of the two primary rhythms that create it analogous to the relationship of the beat frequency to the frequencies of the pitches that create it?

It turns out that the formula of subtracting the frequencies to obtain the beat frequency holds with polyrhythms in some cases and not in others. For example, $\frac{1}{2} - \frac{1}{3} = \frac{1}{6}$, and $\frac{1}{6}$ is the frequency for the polyrhythm created by 1:2 and 1:3. This formula also works for 1:4 and 1:6, rhythms with common-multiple lengths. It does not work for 1:3 and 1:5, however. This is a curious phenomenon, rich with possibilities for exploration and analysis.

- Patterns for operations on fractions in the polyrhythm chart
  If you study the polyrhythm chart, you can see many interesting patterns that the different polyrhythms make with each other. See if you can find a simple way to add fractions based on the chart. Of course you can just add the fractions mathematically with paper and pencil, but there is a way to get the answer directly from the chart. One system works only for fractions with 1 in the numerator. For example, to add $\frac{1}{2} + \frac{1}{3}$ the denominator of the sum is clearly the phrase length of the polyrhythm created by 1:2 and 1:3 (the LCM of 2 and 3). The numerator of the sum is found by adding the phrase lengths of the two rhythms, 2 + 3, so $\frac{1}{2} + \frac{1}{3} = \frac{5}{6}$. Does this system work for fractions with a number other than 1 in the numerator? Why or why not? Could you find a system in the chart that would work for all fractions? What would it be?

- Class performance
  Have a student bring in a drum machine so that you can play any combination of rhythms at will to find a correlation between the sound and the mathematical relationship.

- A brick-laying problem
  Problem 4 of Polyrhythms in Music is similar to a classic brick-laying problem: Two layers of bricks are laid, with each layer constructed from bricks that alternate in length. What is the minimum length the wall can be using specified brick lengths if the brick layers must match perfectly at the end of the wall?

Generalize the problem for alternating lengths in the first layer of a and b and in the second layer of c and d. As mentioned earlier, for some cases the solution is $ad + bc$.

Extension: Establish a rule for all cases, or determine inequalities for a, b, c, and d where particular formulas will work.

ANSWERS

Polyrhythm Chart and Problems

1.

Table 1	
Polyrhythm	**Phrase length**
2:3	6
3:4	12
3:5	15
2:3:5	30

2. In these cases, the phrase length of the polyrhythm is the product of the phrase lengths of the individual rhythms.

3.

Table 2	
Polyrhythm	Phrase length
2:4	4
2:6	6
2:6	6
2:3:4	12

4. No; the phrase length is less than the product of the individual rhythms. On the polyrhythm chart the first rhythm is a divisor of the last.

5. Polyrhythm numbers in Table 1 are not divisible by each other and have no common factors other than 1. In Table 2, the last number is divisible by the first.

6. The phrase length of the polyrhythm is the least common multiple (LCM) of the phrase lengths of the individual rhythms.

7. You have to find the LCM of the phrase lengths of the individual rhythms that make up the polyrhythm. For any group of numbers or expressions, the LCM is the product of the greatest powers of the prime factors of the numbers or expressions.

Polyrhythms in Music

1. The LCM of 13 and 15 is 195.
2. The LCM of 2, 5, and 4 is 20.
3. The LCM of 2, 3, 4, 5, 6, and 7 is 420.
4. 22 beats. Use a graphic approach. Mark segment lengths, alternating 2 and 3 on one line and 4 and 5 on another line. They match on the twenty-second beat.

THE WORLD IS A DRUMMER

Every day the sun rises and sets. This cycle has been repeating since the earth began and will probably go on for some time. Most animals follow the sun closely in their sleep patterns. Other things in our lives—and in the world at large for that matter—run in cycles. Long cycles like the revolution of the earth around the sun determine the seasons. Short cycles like the beating of a bee's wings are barely noticed; they just sound like a buzz. Some cycles are very important to us, such as our heartbeat, a repeating cycle we can both hear and feel.

When you start thinking about it, you realize that thousands upon thousands of things going on around us are running in cycles. Some are made by humans and some are made by nature: the tick of a clock, the turning of a bicycle wheel, the revolutions of a motor, the drip of a leaky faucet, waves hitting the beach, tides, the migration of birds, the return of a comet, and so on.

What does this have to do with mathematics? Well, one thing humans like to do is count things. Have you ever counted the tiles on the ceiling just because you were bored, or counted the cracks in the sidewalk as you walked to school? Most of us enjoy counting money. We always seem to want to know "how much" or "how many." So it's no surprise that we find ourselves counting cycles also. But counting cycles can be kind of messy. Each cycle follows its own pattern with its own length of time before it repeats, and sometimes the amount

of time it takes to repeat a pattern changes from cycle to cycle. As if that weren't messy enough, some cycles happen along with other cycles, for example, waves hitting the beach and the daily tides.

Musicians refer to the rate of a cycle repeating as its *tempo*. Tempo is measured by different things for different cycles, but it is always related to the number of repetitions within a certain amount of time. The tempo of the earth's rotation is one rotation per 24 hours. The tempo of a heartbeat is about one beat per second (or faster when you are exercising). Musicians refer to *rhythm* as the pattern of soft and loud (accented) beats within that tempo. The rhythm of the heart is a loud thump, a soft thump, then a pause.

An important part of what musicians do is count cycles and create rhythms to delight our senses. They are modeling the cycles and rhythms of nature. How counting and combining cycles can create rhythms that seem to provoke emotional responses in people is pretty mysterious.

Let's review what's really going on. Nature contains many cycles repeating at various tempos and creating rhythms; people like to count; people count cycles; musicians count cycles and create rhythms with sounds that remind us of the cycles of nature. This is getting complicated! It's no wonder that every step of the way we're doing mathematics to keep track of it all.

HELP AND INFORMATION

Polyrhythms means many rhythms (just as a *polygon* has many sides). You will learn that the essence of what makes polyrhythms interesting to listen to comes from a mathematical principle we apply to numbers and algebraic expressions.

Special terms and definitions

- *Beat:* An even pulse in time with no accents
- *Rhythm:* A pattern of accents in relation to the beat
- *Polyrhythm:* A rhythm created by a combination of rhythms
- *Phrase length:* The number of beats for a rhythm or polyrhythm to repeat

What to do

Following the direction of your teacher, you will clap to the metronome beat the rhythms assigned to your group. On the polyrhythm chart, each rhythm is represented on a different horizontal row. The vertical lines represent the beats. After each rhythm is clapped, use the polyrhythm chart to make a visual record of the claps. Place marks on the appropriate vertical line to indicate the beats on which you and your classmates clapped. Mark each rhythm with a different symbol. Use a circle, a triangle, a square, an **X**, or something else.

Polyrhythm Chart and Problems

Problem 4: Refer to the polyrhythm chart and observe how the individual rhythms that make up the polyrhythms of Table 1 relate to each other. Now observe the polyrhythms of Table 2. What is the same for all the polyrhythms of Table 1 that is different for all the polyrhythms of Table 2?

Problem 5: Compare the tables as you did in problem 4, but do not use the chart. Instead, focus on the relationship between the numbers that make up the polyrhythms. What do they have in common? How are they different?

Problem 7: This problem is asking you to find a general rule for determining the phrase length of any polyrhythm. How do the numbers in the individual rhythms relate to the phrase length of the polyrhythm?

Polyrhythms in Music

Problem 4: This problem is fundamentally different from the others. Feel free to use a chart and map it out. Read the problem carefully. The general rule for this type of polyrhythm is much more complicated than that for the other polyrhythms.

POLYRHYTHM CHART AND PROBLEMS

Fill in the chart below as you clap the rhythms in class.

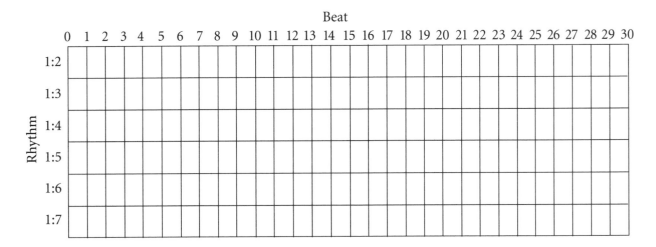

1. Fill in Table 1 for the polyrhythms indicated.

2. Explain in words how the phrase length for the polyrhythms in Table 1 can be determined without clapping or using the polyrhythm chart.

Table 1	
Polyrhythm	**Phrase length**
2:3	
3:4	
3:5	
2:3:5	

3. Refer to the polyrhythm chart and fill in Table 2 for the polyrhythms indicated.

4. Does the mathematical method for finding the phrase length in Table 1 work for the polyrhythms in Table 2? Use the polyrhythm chart to verify your answer, and explain.

5. Compare the polyrhythms in Table 1 with those in Table 2. Look for patterns and explain how the polyrhythm numbers in Table 1 are different from those in Table 2.

Table 2	
Polyrhythm	**Phrase length**
2:4	
3:6	
2:6	
2:3:4	

6. In general, what is a mathematical name that describes the relationship between the phrase lengths of individual rhythms and polyrhythms?

7. Describe a way to find the phrase length for any polyrhythm.

POLYRHYTHMS IN MUSIC

1. *Tabla* drums are commonly played in India. A tabla drummer uses extremely complicated rhythms that can go on for many beats before they repeat. The drummer might play a rhythm that repeats every 13 beats with one hand while simultaneously playing a rhythm that repeats every 15 beats with the other hand. How many beats would it take before the two hands of the Indian drummer were playing together again?

2. Jazz drummers also use complicated polyrhythms in their music. It is common for a jazz drummer to simultaneously play a rhythm in a cycle of 2 with one hand, a cycle of 5 with the other hand, and a cycle of 4 with a foot on the bass drum. How many beats would it take for both of the drummer's hands and the drummer's foot to be striking the drums at the same time?

3. If six drummers were playing different rhythms that accented 2, 3, 4, 5, 6, and 7 beat cycles, how many beats would pass before they were all accenting together? Notice that this is the polyrhythm created by all the rhythms of the polyrhythm chart.

4. Sometimes a drummer will play a pattern that switches back and forth between two rhythms instead of a pattern that repeats just one rhythm. For example, a drummer might accent after two beats, then after three beats, then again after two beats, and so on. If one drummer is playing this pattern and another is alternating accents between four and five beats, what is the smallest number of beats that will pass before the two drummers accent together? Be careful; the answer is not 45! Describe how you determined the answer.

5

Record-Producer Algebra

Using Algebra to Perform Rap Music

TRACKS
29–34

In Record-Producer Algebra, students play the role of a record producer working with a singer, a songwriter, and a band to record a segment of rap music. Producers are responsible for all aspects of the recording process, including writing arrangements or programming computer-sequenced parts that must adhere to requirements set by the artist. Using information about a rap segment, students calculate entrance points and phrase lengths for various vocal parts that have been designated to end at specific places in the musical recording. Students explore guess-and-check strategies and arithmetic solutions of their own design, and they perform vocal renditions of their answers for each problem. Your coaching helps students see algebraic strategies that assign variables, create

equations, and generalize all parameters. The general formulas can be used to alter the existing vocal parts and create new arrangements.

Through this activity, students see how arithmetic solutions that they devise and confidently apply are actually the steps in solving an algebraic equation. They are doing algebra without realizing it. Students are not given a strategy. Instead, they obtain numerical information to solve the problems by listening to the CD tracks and counting the beats. Students express their strategies in words as a transition to algebraic solutions. As the problems become more difficult, students see the usefulness of an algebraic equation as a tool.

Mathematics topics

Counting, algebraic representations, solving linear equations with fractions, creating algebraic models (general formulas), algebraic manipulation of general formulas. *Prerequisites:* Prealgebra-level understanding of variables and equations.

Music topics

Beats, time, measures, pop/rock song arranging, vocal phrasing, rap vocal performance. *No prerequisites.*

Use with the primary curriculum

- During the study of algebraic word problems
  Record-Producer Algebra can be a valuable enrichment activity during units dealing with the translation of real-world scenarios into algebraic models.

- As enrichment for algebraic symbol-manipulation techniques for solving equations
  This activity provides an interesting application of equations and gives practice in solving them.

- Between units in any mathematics course
  Developing algebraic skill and understanding is an ongoing process that can be practiced in any mathematics course. This activity will enrich algebra skills through a different kind of learning experience. Variety in instruction can keep all students interested and involved.

Objectives

- To expand awareness of the use of algebra
- To foster understanding of the connection between arithmetic solutions and algebra
- To use algebraic formulas to solve real-world problems
- To provide opportunities for practicing algebraic manipulations
- To engage students in a fun, unique classroom experience

Student handouts

- Algebra: Everyone's Doing It (reading; one per student)
- The Producer's Problems (resource page; one per group)
- The Producer's Workstation (worksheet; one per student)
- Music/Rap Charts (worksheet—optional; two per group)

Materials

- CD tracks 29–34
- Blank music manuscript paper (optional)
- Overhead transparency of Music/Rap Charts

Instructional time

40–80 minutes

Instructional format

Because students are creating algebraic representations from their own arithmetic solutions, the exercises follow a constructivist format that helps students build on their own understanding. Students create strategies, make connections, and solve problems with minimal teacher intervention. The students' work is then mined by you for algebraic content, and you reveal this content to the students to inform their subsequent work. The activity alternates between students working in pairs or small groups and the whole class coming together to listen to the CD track. The group sessions of listening and discussing limit students' freedom to work at their own pace, because all students must complete the same problem within a short time. In a sense, the classroom flow oscillates between whole-class, group experience (the norm for a band class) and independent problem solving.

It can be fun to assign students the task of creating their own lyrics for Problems 1, 2, and 4 before class. If you choose to do this, you will have to specify the number of syllables in order for each problem to work as written mathematically. (Problem 1 needs seven syllables, Problem 2 needs three syllables, and Problem 4 needs two syllables.) Though this can mean more work, you may have one student (linguistically gifted but mathematically challenged) who would love to have the opportunity to contribute with confidence in mathematics class.

The music CD includes tracks that demonstrate how the rap problems sound when they are performed correctly. You may want to listen to them as you prepare to teach the activity, but don't play them for the class unless your students are having trouble and need to hear a sample. Track 31 demonstrates a correct performance of Problem 1; track 32 is for Problem 2; track 33 is for Problem 3; and track 34 is for Problem 4.

Student preparation

Many students need some time to warm up psychologically to the idea of singing and counting beats in a mathematics class. Give them the reading, Algebra: Everyone's Doing It, a day before the activity, and spend five minutes discussing the ideas it presents. If this is not possible, at least give students some idea of what you will be doing on activity day. The following discussion point gives more information about Leibniz, who is mentioned in the reading.

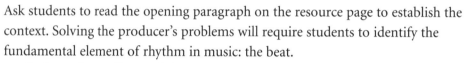 Your students may be interested in learning that Gottfried Wilhelm Leibniz, a scholar of law and philosophy and a career diplomat, used his spare time to explore mathematics. His studies of infinite series led him to the basic ideas of calculus. Sir Isaac Newton had discovered these ideas a few years earlier, but both men developed the ideas independently. Leibniz was the first to publish the ideas of differential calculus in 1684.

ACTIVITY SCRIPT

STEP 1   Set the context and establish the concept of beat

Ask students to read the opening paragraph on the resource page to establish the context. Solving the producer's problems will require students to identify the fundamental element of rhythm in music: the beat.

Refer to the reading and discuss how musicians count beats. Solicit explanations and experiences from musicians in the class. Ultimately, you need to establish that the beat is the fundamental pulse used to measure where all of the events in a piece of music occur. Play CD track 29 to demonstrate where the beat falls in the rhythm that students will rap to.

The CD track marks only the location of the beats; it does not even imply assigning numbers to the beats. Establishing the need to count beats should come from students as an outgrowth of their need to solve the problems.

 Musical students in your class may comment that beats are counted and organized into measures. Measures are a means by which beats are grouped to help organize the structure of music as well as to suggest a shape to the rhythm. The music example that is the basis for this activity is in four-four time; that is, there are four beats per measure, and each beat is represented by a quarter note. It is not necessary for students to grasp the concept of measures for this activity. It can be explored in a discussion, but you would do better to wait until after the first problem so students have the chance to develop methods of counting beats on their own before they are given any instruction.

STEP 2   Problem 1: Determining the entry point for a rap phrase

This problem is an example of a typical calculation a producer or an arranger makes when programming a sequenced production of a rap song or when cuing a singer to enter the recording. The part to be rapped is the seven-syllable lyric line "Now we want to take you there." The writer wants each syllable to fall exactly on consecutive beats with the last syllable occurring on the same beat on which the band enters, as heard on the CD track. The problem for students is to determine on which beat to start the rap so that this will be accomplished.

Introduce this problem with minimal instruction. The problem statements are on the resource page, The Producer's Problems. After students read through the problem description, play CD track 30.

Many students will try to solve the problem by feeling it out or guessing before resorting to calculations. Don't discourage this. When they get frustrated with guessing, they will move to counting and calculating. It is valuable for students to discover this necessity on their own.

Tell students you will play the CD track as many times as necessary. Encourage them to try to rap it correctly and to have fun with it. There will always be a few laughs and great opportunities for closet performers (and the not-so-closet performers) to get some attention. Students can put their scratch work in the space for Problem 1a on the worksheet.

You can expect to see students use many different strategies. Some will make tally marks to help them count the beats, while others will count mentally. Very few will use an algebraic equation on this simple problem. At this point, do not encourage any particular strategy.

Much of the music heard in the media and in popular formats is performed by computers using sequencing software and digitally sampled instrument sounds. The equipment for such productions uses a system called Musical Instrument Digital Interface, known as *MIDI*. When producers program music for MIDI, they often have to make calculations for the instrument and the vocal parts as well as the overall arrangements. Such calculations are similar to those used in solving the problems in this activity.

When the rap has been performed correctly, ask students to share how they solved the problem. Some students will probably be unable to perform the solution, but be sure that all students understand the solution and answer Problem 1b.

STEP 3   Problem 1c: An algebraic solution

While most of your students will be able to do the arithmetic to solve the problem, not as many will be able to easily represent it as an algebraic equation. To help students see the connection, translate their written explanations to an algebraic sentence (equation). It is valuable to point out to students that they often apply algebraic concepts in mental mathematics without realizing it. This problem is a perfect example. Ideally, you can have a student put his or her explanation on the board, and you can then translate it to algebra. You can use the following analysis to demonstrate the solution:

$$\underbrace{\text{beats before rap}}_{x} \quad + \quad \underbrace{\text{beats of rap}}_{7} \quad = \quad \underbrace{\text{total beats}}_{25}$$

$$\begin{array}{ccccc} x & + & 7 & = & 25 \\ & & -7 & & -7 \\ & & x & = & 18 \end{array}$$

Most students will automatically perform the arithmetic on the right side of the equation, $25 - 7 = 18$, without using a variable. The same students can be mystified by the idea of subtracting 7 from both sides of an equation. The above analysis can show them the connection. This simple example does not make a strong case for using an equation with a variable because the arithmetic solution was obvious. Urge students to withhold judgment on the usefulness of equations—it will become more apparent when they encounter more difficult problems. Ultimately, try to bring students to see that the arithmetic solutions they use to solve the problem are in fact the steps used to solve the algebraic equation.

Shown on the following page is the standard musical notation used in marking rhythm. Notation of this sort is often referred to as a *chart*. (A chart in jazz and pop music refers to written musical parts for musicians that often contain chord symbols with rhythmic notation, combined with notated melodies.) It is similar to what you may see some of your students doing on their own (making tally marks). In general, music is written on five parallel lines called a *staff*. The lines and spaces of a staff correspond to various musical notes. In contexts like those in this activity, which involve rhythm but not notes, slash marks are used on the staff as shown. Each slash mark corresponds to a beat, and each vertical line (bar line) groups the beats into measures.

Use the blank Music/Rap Charts from the student worksheet on an overhead transparency as you analyze the beats with your class. The chart that follows shows how Problem 1 would be notated.

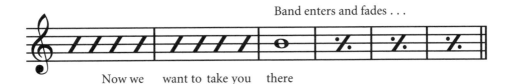

Band enters and fades . . .

Now we   want to take you   there

Students who made tally marks on their scratch paper will be excited to see that what they did on their own is very similar to how professional musicians write this type of music. Bring this to their attention.

You will need to decide whether or not to provide students with the Music/Rap Charts to use as worksheets. If students are having difficulty with the problems or have minimal experience with variables and equations, the charts may be necessary. But be aware that providing the charts will decrease students' reliance on using equations and calculations to solve the problems.

STEP 4   Problem 1d: Generalizing for rap length *r*

Allow students to work on this problem independently in groups. When most have come up with an answer, ask them to share their solutions.

Some students may be unclear as to what variable to solve for. The problem is purposely worded in terms of its application intent—to provide information on how to cue a singer to enter. Students need to make the connection that for this to happen they need to set up the formula so it yields the number of beats before the rap, *x*. If some groups wonder what variable to solve for, try to coach them through this reasoning rather than telling them that they need to express *x* in terms of *r*. If some groups have other difficulties with the generalization, point out that they can refer to the previous problem and simply use *r* instead of 7 and then solve for *x*.

An analysis to demonstrate the solution to students is shown below.

$$
\begin{array}{ccccc}
\underbrace{\quad x \quad} & + & \underbrace{\quad r \quad} & = & \underbrace{\quad 25 \quad} \\
 & & -r & & -r \\
 & & x & = & 25 - r
\end{array}
$$

T E A C H E R   N O T E S

STEP 5   Problem 2: Phrase made from repeating a rap

The remaining problems follow a process similar to that outlined in Steps 2–4.

Direct students to read Problem 2 on the resource page. After they have read it, play CD track 30. When students are able to perform the part, have them work independently to finish the remaining parts of Problem 2 on their worksheets. Have students share their results with the class. A musical chart of the rap part, shown below, can be used to examine the problem and connect students' counting to the mathematical aspects.

take me there

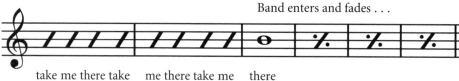

Band enters and fades . . .

take me there take   me there take me   there

As for Problem 1, you can use the following analysis to demonstrate how the arithmetic solutions students use to solve the problem are in fact the steps used to solve the algebraic equation representing the problem.

$$\underbrace{\text{beats before pattern}}\ +\ \overbrace{\qquad\qquad\text{beats of pattern}\qquad\qquad}\ =\ \underbrace{\text{total beats}}$$

$$\underbrace{\text{beats before pattern}}\ +\ \underbrace{(\text{number of repetitions})}\underbrace{(\text{beats of rap})}\ =\ \underbrace{\text{total beats}}$$

$$x\ +\ (4)\qquad\quad(3)\ =\ 25$$

$$x\ +\ 12\ =\ 25$$

$$-12\qquad\qquad -12$$

$$x\ =\ 13$$

STEP 6   Problem 3: Finding a phrase length

Ask students to read Problem 3; then discuss it with the whole class so that all students understand what is being asked. Students will first need to calculate (no guessing) before they can try to perform, since the words are not written and they do not even know what to sing yet. They must calculate the length of a rap phrase (number of syllables) that, when repeated three times, will start with the bass guitar on the third beat of the third measure and end where the band enters. In light of this, the worksheet directs students to an algebraic solution as their first step. Play the CD track and encourage students to use a musical chart to aid them. The algebraic analysis appears on the top of the next page.

$$\underbrace{\text{beats before bass guitar}}_{} + \overbrace{\text{beats of pattern}}^{} = \text{total beats}$$

$$\underbrace{\text{beats before bass guitar}}_{} + \overbrace{\underbrace{(\text{number of repetitions})}_{}\underbrace{(\text{beats of rap})}_{}}^{} = \underbrace{\text{total beats}}_{}$$

10	+	(3)	(r)	=	25
10	+		3r	=	25
− 10					− 10
			3r	=	15
				r =	5

Many students will count beats from where the bass enters to determine the total number of beats. This would simplify the solution. You might mention that, while this method is easier, counting from the beginning of the section would be necessary in programming computer-sequenced parts.

After students have calculated the number of syllables for the phrase, they need to write some words. You can have fun with this, but it can also take the class off on a tangent. You might suggest to students that they consider where the phrases overlap with the previous parts and write words so that some will be in sync. Two suggestions that honor this to varying degrees are "Can I take you there?" (three words overlap with the first phrase) and "Want to take you there" (all words overlap with the first phrase).

If you have time, you can have several groups of students try several versions of the rap. Don't allow this to distract from the mathematical focus of the activity. If you find the writing of words to be too much of a distraction, assign students the suggested phrases.

The musical chart below uses the phrase "Can I take you there?"

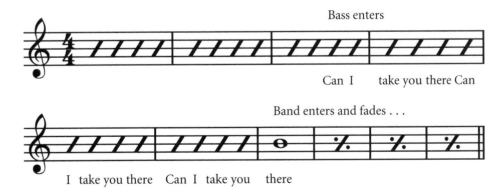

T E A C H E R   N O T E S

After students have performed the parts with the CD, let student groups finish the generalizations (Problem 3c–e) on their own. Before moving on to Problem 4, share solutions and analysis with the class, using a process similar to that outlined for the previous problems.

You might consider saving Problem 3e for a homework assignment, completing it at the end of the activity, or even skipping it altogether. If students' skill level with algebraic symbol manipulation is low, time spent on this problem can disrupt the momentum of the activity, and it is best saved for later. On the other hand, if students can solve the other problems with little difficulty, this problem can add richness and variety to the interactive exercises.

STEP 7   Problem 4 (optional): Using space between words

Problem 4 can be omitted or addressed during a subsequent class session, depending on how quickly your students have moved through the previous exercises.

This problem is different from the others in that the phrase considered contains some beats with no words. It is a four-beat phrase consisting of two words followed by two empty beats. The empty beats are referred to in music as *rests* (see Measures of Time, Part I). Since the requirements of the songwriter are that the end of the phrase must be a rapped word and the phrase as defined should end with two beats of rest, the five repetitions of the phrase must actually be four and a half repetitions, or five repetitions minus the last two beats of rest. This gives rise to various interpretations for an algebraic solution, and in any event a more complicated one. Your students will probably find it easier to work with the musical chart or with arithmetic to find the solution.

Play CD track 30 and have students rap their solutions. Following are an algebraic solution and a musical chart for you to use while discussing the problem with students.

beats before pattern +	beats of pattern		= total beats
beats before pattern +	(beats of phrase)(repetitions)	− rest beats	= total beats
beats before pattern +	(word beats + rest beats)(repetitions)	− 2 rest beats	= total beats

$$x \quad + \quad (2 \quad + \quad 2) \quad (5) \quad - \quad 2 \quad = \quad 25$$
$$x \quad + \quad 18 \quad = \quad 25$$
$$-18 \qquad\qquad -18$$
$$x \; = \; 7$$

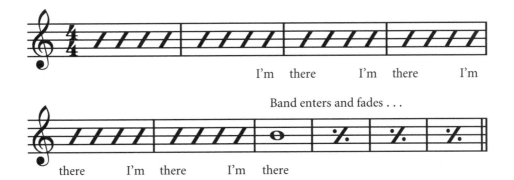

I'm   there   I'm   there   I'm

Band enters and fades . . .

there   I'm   there   I'm   there

STEP 8   No problem! Performance and analysis of the rap parts

Divide the class into four performance groups and assign each group a different rap part from Problems 1–4. Play the CD track and have students perform all the raps simultaneously. Experiment with performing different combinations of the parts together to find the combinations that are the most interesting or pleasing to the class. Most students will have to refer to their solutions and count in order to perform the parts successfully.

If you have time, hold a class discussion on the written analysis questions in the Follow-up Activities. If you do not have time for discussion and you plan to have students do a journal exercise or an extension, after the performance have them take notes on their observations of how the parts seemed to work together, making judgments on which combinations they liked best and why.

FOLLOW-UP ACTIVITIES

Written analysis

Have students use their discussion notes and problem solutions to respond in a written analysis to the following prompts:

- Which combinations of raps did you think sounded best together? Explain why.

- What could be changed in the other combinations to make them sound better? Explain why in terms of their mathematical relationships.

- Which of the formulas that you created could be most useful in helping you work out your proposed changes and perform them?

TEACHER NOTES

Writing prompts

Either instead of or in addition to the written analysis, it is valuable to have students reflect on their learning process and on some of the affective issues raised in the activity. The following prompts are suggested:

- What did you learn in this activity?
- What was the most interesting or valuable thing you learned?
- Was the activity fun?
- Did you realize before doing this activity that rock and rap music production involves algebra?

Homework assignments

Follow Record-Producer Algebra with practice solving equations. The often dry and tedious algorithms will have a fresh sense of relevance. Use exercises that involve creating algebraic models, solving linear equations, or solving general formulas for specified variables or word problems that emphasize writing equations.

Projects

- Make the changes you suggested in the written analysis and perform them for the class. Show how you used the formulas and which formulas are useless and why.
- Interview a real record producer and find out what kinds of mathematical problems producers actually face. Present your findings to the class.

ANSWERS

The Producer's Workstation

1b. Answers will vary. Example: "We counted all the beats that there were before the band came in and came up with 25 beats. Since the phrase has 7 syllables, we subtracted 7 from 25 and got 18. We knew we had to start the rap the beat after the eighteenth beat."

1c. $x + 7 = 25$

$x = 18$. The singer must enter on beat 19.

1d. $x + r = 25$

$x = 25 - r$

2b. Answers will vary. Example: "Since the phrase has 3 syllables and needs to be repeated 4 times, we multiplied 4 times 3 and subtracted that from 25. This gave 13, so we started rapping on beat 14."

2c. $x + 4(3) = 25$
$x = 13$

2d. $x + n(3) = 25$
$x = 25 - 3n$

3a. $10 + 3r = 25$
$r = 5$

3b. Answers will vary. Suggestion: "Where are we going?"

3c. $10 + nr = 25$
$r = \dfrac{15}{n}$

3d. $x + nr = 25$
$r = \dfrac{(25 - x)}{n}$

3e. $x = B - nr \qquad n = \dfrac{B - x}{r}$

$\quad\; r = \dfrac{B - x}{n} \qquad B = x + nr$

4b. $x + (2 + 2)(5) - 2 = 25$
$x = 7$

4c. $x + (r + s)(5) - s = 25$
$x = 25 - 5r - 4s$

4d. $x = B - (r + s)n + s \qquad n = \dfrac{B + s - x}{r + s} \qquad r = \dfrac{B - x - s(n - 1)}{n}$

$\quad\;\; B = x + (r + s)n - s \qquad s = \dfrac{B - x - rn}{n - 1}$

ALGEBRA: EVERYONE'S DOING IT

A lot of what you learn to do in algebra class you have been doing since third grade. That's right, you have been doing algebra since the time you started to count and figure out simple calculations such as how much more time you had to wait until your birthday or how long it would be until class was over. (Let's see . . . I have been here for 20 minutes and the class is 50 minutes long, so there are 30 minutes left; $50 - 20 = 30$.) Record producers also think in this way when they communicate with writers, singers, and musicians.

Record producers are responsible for overseeing all the decisions that take place in recording a song. They help the musicians and singers understand what the songwriter wants, decide which takes (recorded samples) should be kept and which need to be redone, decide how to make the sound of the instruments complement the vision of the artists and the style of the song, keep everyone happy—the list goes on and on. With all the different responsibilities a producer has and the variety of talents the job demands, are you surprised that record producers use algebra to produce hit songs? It's true, even though they may not realize it! Let's take a look at what's going on.

Musicians organize music according to something called the *beat,* a reference point that musicians use to count time in music. By counting beats, musicians know when to play or not play and

they keep track of the number of beats that pass before their part comes in. So, as you can imagine, when musicians play music they are always counting in some form or another, depending on how much they actually have to think about it. There are convenient ways to group beats together so counting is easier, sort of like you do when you buy three dozen eggs. Since eggs are packaged in groups of 12, when you buy three cartons you don't have to count to know you have 36 eggs. You can just multiply 12 by 3.

As people get used to working with music, they don't need to count at all (at least not the way you're used to). Their counting almost becomes unconscious. The seventeenth-century mathematician Gottfried Wilhelm Leibniz had a great way of expressing this: "Music is the pleasure the human soul experiences from counting without being aware it is counting." Think about that—doing mathematics without even being aware of it! You may find it hard to believe, but you do this all the time, and you have been doing it since long before you started learning mathematics in school.

In this activity you will arrange a pop rap tune by counting, doing calculations, and writing algebraic formulas in ways similar to the methods writers, musicians, and record producers use. You just might discover yourself doing algebra without even realizing it!

THE PRODUCER'S PROBLEMS

In today's activity you are playing the role of a record producer working on the timing for a rap song. For each problem, your teacher will play a section of the song. Listen to it and work through the problems as described on the worksheet.

Problem 1

The songwriter wants a rap phrase to end exactly where the band comes in on the CD track. The phrase is "Now we want to take you there." The songwriter is very picky and has these requirements for how the phrase should be sung:

- The syllables of the phrase must fall exactly on consecutive beats. There can be no words between beats and no beats between words.

- The last syllable of the phrase must fall on the same beat on which the band enters.

Where should the singer start the rap so that the requirements are met?

Problem 2

The writer wants to add another voice with the phrase "Take me there." The phrase is to be repeated four times, and the last syllable of the last repetition must land on the beat on which the band enters.

Where should the singer start the rap so that this is accomplished?

Problem 3

The songwriter would like another rap line to enter on the same beat on which the bass guitar enters and he would like this rap line to repeat three times. As with the raps before it, he wants the syllables to fall on consecutive beats with the last syllable falling on the beat on which the band enters. To write the words he needs to know how many syllables the phrase must have.

How many syllables must the rap phrase have?

Problem 4

To add more variety, the songwriter wants one last phrase to be sung. This phrase is different from the others in that it does not immediately repeat. The songwriter wants the phrase "I'm there" to be followed by two empty beats. It is sung five times and ends so that the word *there* of the last cycle lands on the same beat on which the band enters.

Where should the singer start the rap so that this is accomplished?

Problem 5

No problem! Perform all the rap parts together as a class. If you don't like how the parts fit together, you can change them and use the formulas you created to help the class perform them correctly.

THE PRODUCER'S WORKSTATION

Problem 1

a. Volunteer to rap the phrase with the CD track when you feel confident you can do it correctly. Show your scratch work, which might include sketches, calculations, counting, or other notations.

b. Explain clearly in words how you figured out when to begin the rap phrase.

c. Solve the problem using an algebraic equation. Let x represent the number of beats before the singer enters.

d. Generalize the problem and write a formula for how you could direct a singer to enter for any length rap phrase, r.

Problem 2

a. Volunteer to rap the phrase with the CD track when you feel confident you can do it correctly. Show your scratch work.

b. Explain clearly in words how you figured out when to begin the first phrase.

The Producer's Workstation (continued)

    c. Solve the problem using an algebraic equation. Let *x* represent the number of beats before the singer enters.

    d. Generalize the problem and write a formula for how you could direct a singer to enter for any number of repetitions, *n*, of the same phrase. Let *x* be the number of beats before the singer enters.

Problem 3

    a. Use an algebraic equation to solve this problem. Let *r* represent the length of the rap phrase.

    b. After you have found the number of syllables for the phrase, make up some words for the phrase and volunteer to rap it with the CD track. Write your words below.

    c. Generalize the problem and write a formula for the length, *r*, of a rap phrase repeated *n* times.

    d. Generalize the problem further and write a formula for the length, *r*, of a rap phrase, letting *x* represent the number of beats before the rap starts and *n* the number of repetitions.

    e. Generalize the problem completely and write a formula for *r* if the entire CD track before the band enters is *B* beats. Create three more formulas by solving for *x*, *n*, and *B*.

The Producer's Workstation (continued)

Problem 4

a. Volunteer to rap the phrase with the CD track when you feel confident you can do it correctly. Show your scratch work.

b. Show your solution as an algebraic equation. Let x represent the number of beats before the rap should start.

c. Generalize the problem and write a formula for how to direct a singer to enter, letting r represent the number of syllables of the rap, s the number of empty beats between the raps, and x the number of beats before the singer enters.

d. Generalize this problem completely, letting B be the total beats before the band enters and n the number of repetitions of the phrase. Write five different formulas, each solved for a different variable: x, B, s, n, and r.

MUSIC/RAP CHARTS

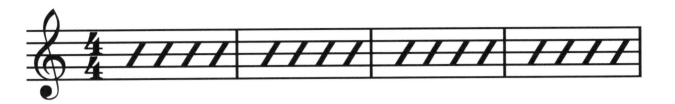

Band enters and fades . . .

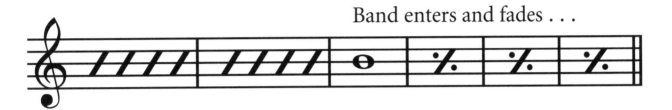

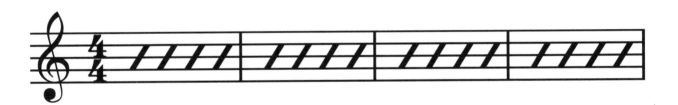

Band enters and fades . . .

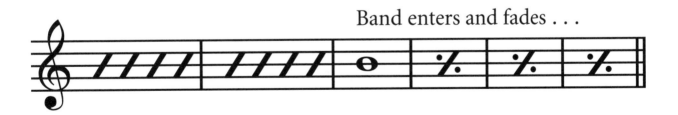

6

Functional Composer, First Movement

TRACKS
35–40

A Mathematical Solution to Writer's Block

In Functional Composer, First Movement, students use transformations of an algebraic function to solve a music composer's problem. They represent a musical motif as a function, calculate the function's transformations, draw and see the graph of those transformations, and hear the transformations played as melodic lines. The qualities that relate the transformations of the function to one another mathematically are present in their melodic counterparts, so students experience these mathematical relationships sonically, in addition to numerically and visually. They discover that these relations are the glue of a musical composition that has structural integrity and hence sounds good and makes sense aesthetically. The activity closes with students identifying the melodic counterparts of the transformations in a musical composition. The composer's problem is solved.

Mathematics topics

Definition of a function, function notation, graphing transformations of functions (stretching, shrinking, reflecting, translating), composite functions, order of operations, in/out tables. *Prerequisites:* Ability to graph ordered pairs of numbers on a Cartesian coordinate system using in/out tables.

Music topics

Musical notation (pitch), musical composition and melodic structure, ear training, transposition, modes. *No prerequisites.*

Use with the primary curriculum

- To introduce the definition of a function
  Use Functional Composer to introduce the notion of a function as a pairing of integer sets and to provide practice using in/out tables to graph on a Cartesian coordinate system.

- To introduce transformations of functions and their graphs
  Use Functional Composer to present the idea of transforming a function when interpreting and creating graphs of conic sections or logarithmic, trigonometric, or algebraic functions.

- To review functions and transformations
  After the idea of function or transformations has been taught, use Functional Composer to show an application and enrich student understanding.

- To illustrate transformations in a geometry class
  Functional Composer can reinforce student understanding of mappings, dilations, and reflections.

Objectives

- To understand functions and their graphs
  The experience of hearing, calculating, and creating visual representations enhances understanding for all students and gives audio and visual learners a chance to shine.

- To retain the concept of a function
  Retention will be increased by the very memorable experiences of Functional Composer, First Movement.

- To observe unity, elegance, and beauty
  Seeing an artistic problem solved with mathematics stimulates questions about how mathematics relates to nonmathematical realms of intuition and inspiration.

Student handouts

- A Composer's Problem (reading; one per student)
- What to Do (resource page; one per group)
- Melodic Variations and Transformations (worksheet; one per pair)
- Melodic-Contour Graph (worksheet; one per pair)

Materials

- CD tracks 35–40
- Overhead transparencies of student worksheets

Instructional time

50–60 minutes

Instructional format

You conduct the activity by using the CD tracks (pausing and repeating tracks when necessary) together with overhead displays of the transformations. Students work in pairs. One student in each pair uses the Melodic Variations and Transformations worksheet to calculate the function transformation values (using in/out tables) and to write corresponding melodies. The other student uses the Melodic-Contour Graph to graph the various melodic variations (function transformations). It is a good idea to ask students to make notes after each transformation to summarize the nature of the transformation. Try to have your students supply the conclusions stated in the activity script. This process is continued in Functional Composer, Second Movement. Students can make a table in their notes similar to the chart for that activity.

The melodic variations as played on the CD are given a rhythmic shape to enhance their appeal. This rhythmic element has been excluded from the musical notation of the motif to focus attention on the melodic/function aspects.

 Instead of using the prerecorded CD, ask a student to bring a musical instrument to class to play the melodies developed in the lesson. You will still want to use the CD to play the final composition created from the transformations. Musically inclined students can greatly enhance the engagement of all students in class discussion, and even allow students to create their own compositions from the transformations.

Student preparation

To save time on the day of the activity, start the day before with A Composer's Problem (read by students or read aloud by you). Discuss the reading and ask musicians in the class to comment on their own experience.

ACTIVITY SCRIPT

STEP 1   Establish the context

Set the context with students by referring to the reading or describing a hypothetical scenario of a composer who is experiencing writer's block.

> A composer is trying to write a piece of music. She has been inspired with a little melody (motif) but can't seem to come up with any more music that sounds good or makes sense with the melody. Everything she tries to add sounds unrelated to the original melody. She is very frustrated; she is experiencing a bad case of writer's block.

Then suggest that there is a mathematical solution to her problem.

 It can be fun to dramatize the opening scenario and ask students to share their own experiences with writer's block and why it happens. Students may have experienced blocks in other art forms, such as painting or poetry.

 The process used in Functional Composer, First Movement has been used consciously and unconsciously in composition and improvisation by musicians and composers from Johann Sebastian Bach to John Coltrane. Bach's fugue and canon forms were assembled from vast assortments of melodic variations similar to the function transformations we will develop. Jazz improvisers invest great effort in creating *riff patterns* that they use as resources for improvisation. Many of these patterns are generated using a process identical to mathematical transformations.

STEP 2   Establish the number/note system

Turn students' attention to the number/note reference key on the Melodic Variations and Transformations worksheet. Explain that musical notes are different frequencies of vibration that our ear hears as different *pitches*. You can mention that musicians name the different pitches with letters and that the pitches are arranged in a *scale*. The starting point for the musical scale is a note called the *key note* or *tonic* of the scale.

Students will give each note on the scale a number, letting the key note be number 1. The number/note reference for the scale is like a number line. High notes (fast vibrations) are named with higher numbers, and low notes (slow vibrations) are named with lower numbers.

Point out that musicians represent the different pitches by using dots on a set of lines and spaces called a *staff*. Each position of a dot on the staff corresponds to a different pitch. Take time to examine the number/note reference key and point out that the dots fall on either a line or a space. For Functional Composer, First Movement, students need only make a connection between the number and the position of the note on the staff.

Use your own judgment as to how many musical terms you discuss. It is not essential that students know these terms for the success of Functional Composer, First Movement, and too much information can be a distraction. If you do decide to use many musical terms, you might want to keep a word list of new terminology. The letter names of notes need not be mentioned, as this can distract from the goal of the exercise—that students make the geometric observation of where each number's note is positioned on the staff.

STEP 3   Introduce the original motif: $y = f(x)$

TRACK
35

Play the recording of the composer's original motif, and show students that it consists of the following sequence of notes: 3, 5, 2, 4, 3, 0, 1. Play this melody several times if you like.

If your students are musically inclined, you might ask them to determine the numerical values of the notes solely from listening to the melody. You could even ask students to sing the numbers with the melody.

The number values for the notes of the melody will be the function values of $f(x)$. The input values, x, are the numbers of the notes in sequential order: 1 for the first note, 2 for the second, and so on.

Discuss how this fits the definition of a function: the pairing of numbers from two sets, an input value x and an output value $f(x)$. Our motif function is defined as the pairing of numbers between the sequential value of the term, x, and the numerical value of the note, $f(x)$. Students may be uncomfortable because no mathematical rule relates the input values to the output values. A mathematical rule is not essential in the definition of a function, and working with this idea can actually help broaden students' understanding of a function.

It is intriguing to entertain the idea of having no formula to describe the relationship between input and output values. Is this relationship outside the realm of mathematics? What determines it?

One student from each pair should write the first in/out table for this melody and place the correct notes on the staff. Refer to the completed Melodic Variations and Transformations worksheet and Melodic-Contour Graph. (Note that students will complete these worksheets one function transformation at a time.) The other student in each pair will draw the melodic contour on the Melodic-Contour Graph. This motif is the starting point; students will now calculate an assortment of transformations, plot them, and listen to them.

Discuss with students whether connecting the points with line segments on the contour graph is an accurate representation of the function. Have them explain why it is not. Assist them, if necessary, by explaining that the notes do not gradually change from one to the next; instead, the change from one note to the next is like a step. Then point out that they will connect the dots to help them gain a visual image of the shape of the functions/melodies.

To build students' group-work skills, you can have them switch roles for each transformation. This can increase students' interaction but will require more time, since students need to learn both tasks—ideally from each other.

STEP 4 Introduce the first transformation: $y = f(x) + 2$

TRACK 36

Ask students for suggestions of how this function might look and how it might sound. Have them predict how it is related to the original motif, both visually and musically. You might ask a student volunteer to sing the transformation.

Play the melody from the CD after all students have calculated the in/out table, converted the function values to notes, written the melody, and plotted its graph. It is important that students hear the variation (translation) in relation to the original melody so that the specific characteristics of adding a value to a function (translating it in the y direction) become clear.

As you play the function on the CD track, have students follow along the graph on their worksheets or trace the contour of the graph on the overhead in sync with the melody. Ask students to describe or write about the relationship between $f(x)$ and $f(x) + 2$, from both looking at the graph and listening to the melodies.

Mathematics perspective: Adding a value to the function has not changed the shape of the original, only its position on the coordinate grid. The geometric shape of $f(x) + 2$ is congruent to $f(x)$. The transformation can be thought of as either adding 2 to $f(x)$ to get $y = f(x) + 2$ or subtracting 2 from y to get $y - 2 = f(x)$. Thus we can make the observation that subtracting 2 from y moves the function *up* 2 units. (What might adding values to y do to the function?) The idea of values being added to and subtracted from x will be discussed in Functional Composer, Second Movement.

Music perspective: The melody $f(x) + 2$ is within a new scale, called a *mode* of the original scale. This melody in a different mode sounds similar in shape to the original melody, only higher in pitch. Technically the melody created here is not an exact translation of the original melody because the distances between adjacent notes in the major scale is not consistent throughout the scale. Thus the translated melodies will have slightly different qualities from the original, while the mathematical translations will be exactly the same. An exact translation of a melody is called a *transposition.* Turning a translation into a transposition requires adjusting the notes with flats and sharps. Discussing this subtlety only complicates matters and distracts from the message of the lesson, but you should be aware of it in case students bring it up.

Conclude by confirming that this transformation is a *translation* and summarize its characteristics: the congruence of the shapes and the direction of shift.

STEP 5 A second transformation: $y + 2f(x)$

Ask the class to suggest another transformation that could be applied to $f(x)$ that might yield a characteristic different from that of the translation. Lead them to $y = 2f(x)$.

Students may suggest many transformations that will produce function values that cannot be converted to notes as defined by the system in Functional Composer. Function values must be integers, and any function value greater than 14 will be out of the musical range of available notes. Lead students to realize this if they suggest unusable transformations.

Present this and the following transformations using the format of Step 4. As students calculate, plot, and graph each function, play all previous transformations from the CD so that students hear the functions in relation to one another for comparison. Ask students to compare all previous functions to one another and to the original, both visually and musically.

Multiplying a function by a positive integer value stretches the function, changing its shape. It is no longer congruent to the original. What if you multiply the function by a fraction? In some cases this would yield fractional function values that cannot be graphed since $f(x)$ is only defined for integers. Ask students what relationship must exist between the function values and a fraction multiplier to create a graphable transformation. What about multiplying a function by a negative integer? Our fourth transformation will explore the effect of negative numbers.

Geometry connection: Make connections to the topics of geometric mapping: congruence mapping (isometries), reflections, and dilations. Identify which transformations are dilations—those resulting from multiplying the original function by something—as opposed to congruence mappings—those resulting from something being added to or subtracted from the original function.

Conclude by confirming that this transformation has the effect of stretching the original function in the vertical direction.

STEP 6   A third transformation: $y = 2f(x) + 4$

TRACK 38

Again, ask students for another possible way to alter the original function. A student might suggest doing both transformations: multiplying and adding. Lead students to $y = 2f(x) + 4$. Guess and discuss what this transformation will sound and look like and follow the same process as in previous transformations.

The topics of composite functions and/or order of operations can be explored here. What comes first: the stretching or the translation? Is stretching a function by a multiplier of 2 and then translating it by a value of 4 different from translating it by a value of 4 and then stretching it by a value of 2? What is the difference? How would you represent these two processes mathematically? Have students come up with an explanation and the mathematical expressions $y = 2(f(x) + 4)$ (translating before stretching) and $y = 2f(x) + 4$ (stretching before translating). Note that after simplifying, the expressions show a difference. Explain that $y = 2f(x) + 4$ is called a *composite function* because it is a combination of $y = 2f(x)$ and $y = f(x) + 4$. Review of this term can be used as a lead-in the next day for a lesson on composite functions.

Conclude by confirming that multiplying and then adding a value both translates and stretches the function.

STEP 7   A fourth transformation: $y = -f(x)$

Now you can answer any inquiry about multiplying functions by negative numbers. Follow the same process as before as you present this transformation.

Notice that the transformation resulting from a negative multiplier creates a melody that may sound sinister. This can be a thought-provoking connection to make with students.

Conclude by confirming that multiplying a function by a negative value reflects the function across the x-axis.

STEP 8   Does this make music? A composition

Refer back to our unhappy composer with writer's block. Ask students whether the melodies that have been created seem related to one another. Suggest that playing these melodies back to back and adding some accompaniment might be a way to create some real music that sounds good and "makes sense" to our ears.

Track number 40 on the CD is an example of this process. The melodies created from our transformations are assembled with some minimal accompaniment. Some melodies have been repeated and overlapped to create a simple example of how a musical composition might develop from our work. It is strikingly musical and supports our idea that mathematics is a big part of the structure that makes the music aesthetically pleasing.

Play the composition for students and discuss these aspects with them.

Have students identify the functions in the composition as it is played. You may need to play it several times. You might have students go to the overhead and trace the graph of a particular melody as it occurs in the composition.

FOLLOW-UP ACTIVITIES

Functional Composer, Second Movement

The Second Movement explores a question that arises in the First Movement: What is the effect of operating on the input value of the function? In the Second Movement, the original motif is repeated to become a periodic function, and the transformational and musical relationships are revealed in greater depth.

Name That Function

In this activity, students have the opportunity to write the algebraic expression for a mystery melody that is played for them. The activity can be continued throughout the year, perhaps as an occasional warm-up exercise to reinforce graphing concepts in a fun, gamelike format.

Writing prompts

It is valuable for students to reflect on what they have learned. Here are some good writing prompts:

- Did the music created by mathematics sound "good" to you? Explain why or why not.

- Do you think that the mathematics provided a good solution for the composer?

- Do you think that music that comes from inspiration is always better than music that is created mathematically? Explain why or why not.

- Under what circumstances do you think mathematics is an appropriate tool for writing music?

- How do you think mathematics relates to intuition and inspiration?

Textbook assignments

It is helpful to follow Functional Composer, First Movement with textbook assignments that use transformation of functions directly in a purely mathematical context. In this way the mathematical relevance of the activity is reinforced, and the experience of the activity will transfer more effectively to mathematical understanding.

Extensions

- Explore more original transformations and have students bring in musical instruments to play.

- Have students "compose" music algebraically by creating their own transformations and assembling them in different orders. Nonmusician students can create algebraic compositions, and their musician classmates can play them. Working in pairs as composer teams, they can make judgments about the musicality of each transformation, following a guess-and-check process to come up with something that sounds good to them. Is there some commonality within the families of transformations that sounds good? Why? This area is rich with possibilities for exploration.

- Consider including the element of rhythm and applying functions to rhythmic motifs.

ANSWERS

Melodic Variations and Transformations

Original Motif

$y = f(x)$

3   5   2   4   3   0   1

Melodic Variations

$y = f(x) + 2$

5   7   4   6   5   2   3

$y = 2f(x) + 4$

10   14   8   12   10   4   6

$y = 2f(x)$

6   10   4   8   6   0   2

$y = -f(x)$

−3   −5   −2   −4   −3   0   −1

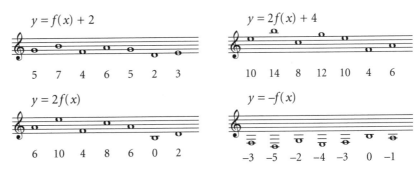

In/Out Table

x	$f(x)$	$f(x) + 2$	$2f(x)$	$2f(x) + 4$	$-f(x)$
1	3	5	6	10	−3
2	5	7	10	14	−5
3	2	4	4	8	−2
4	4	6	8	12	−4
5	3	5	6	10	−3
6	0	2	0	4	0
7	1	3	2	6	−1

Melodic-Contour Graph

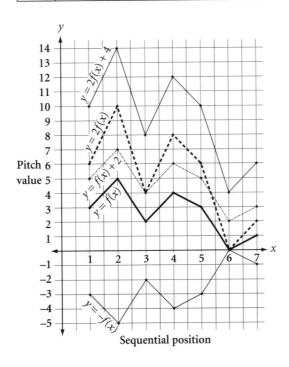

A COMPOSER'S PROBLEM

Did you ever wonder how composers or songwriters come up with their ideas? Where does their music come from? How are they able to make so many melodies go together so well? What holds it all together?

Some songwriters seem to have no idea. They say, "The song just came to me one day," as if they didn't have anything to do with it—they simply heard the song in their head, and all they had to do was write it down. Some composers of symphonic music also say the same thing: that they just hear the music in their head and write it down. Beethoven did this when he was deaf!

As it turns out, not all composers and songwriters are so lucky. Many of them struggle to write music that makes sense and will connect with people emotionally. But what if a composer gets stuck? As you might imagine, this actually happens a lot in real life to creative artists—it is called *writer's block*. Writer's block is when artists aren't inspired and when nothing they write seems to fit or connect well with what they have already written. They're stuck. Can some kind of system or formula help artists get past their writer's block?

There is hope for musicians and composers with writer's block. They can use certain techniques to write music that will work, even if it doesn't

come from some mysterious inspiration. Musicians can think through their knowledge of music theory for help. It may surprise you to know that one technique many composers have used throughout history, and still use, is similar to what you are studying in math class! That's right, mathematics formulas can be used to create good music—even to make musical compositions.

What about those lucky musicians who simply hear music in their heads and write it down? How is that music, created from pure inspiration, different from the music created by theory or formulas? When we look at "inspired" music closely and try to figure out why it works so well, we find that it has some of the same features that our mathematics formula music has! You could say that the mathematical organization of the music is one of the reasons why it is nice to listen to. Amazingly, some composers use mathematical organization without even realizing it.

It seems that no matter where music comes from—whether from mysterious inspiration or from formulas and equations—it has a structure to it that makes it work, a structure that lies in the heart of mathematics. In this activity we are going to use this idea to help a struggling composer overcome writer's block.

WHAT TO DO

A composer is having difficulty writing a piece of music. She has come up with one melody that she likes, but every other melody she thinks of doesn't sound good to her or doesn't seem to go well with the original melody. In today's activity you will use mathematical functions to create some new melodies that could be used to write a piece of music and solve the composer's problem.

Listen to the original motif.

The melody your teacher will play for you is the only melody the composer can think of. We will call this the *original motif*. The number equivalents for the notes of this melody are 3, 5, 2, 4, 3, 0, 1. These note values, when paired with their position in the sequence of notes (1, 2, 3, 4, 5, 6, 7), form seven ordered pairs. These ordered pairs define a function, so we will call our original motif $y = f(x)$.

Record the numbers that correspond to this melody.

Use the number/note reference key on the Melodic Variations and Transformations worksheet to find the numbers that correspond to the notes of the original motif. Place the notes on the staff, and place the number equivalents below the notes. Label the original motif $y = f(x)$.

Make an in/out table.

Now you need to make an in/out table for $y = f(x)$. Let the x-value of the first note be 1, the second note 2, and so on. The note value from the number/note reference key will be $f(x)$. Place the ordered pairs $(x, f(x))$ in the space on your worksheet labeled In/Out Table.

Graph the function.

Plot the ordered pairs of $y = f(x)$ on the Melodic-Contour Graph. This graph shows a picture of the shape of the melody. To see the shape more clearly, connect adjacent points with straight lines.

Create melodic variations using transformations of the function $y = f(x)$.

You are now ready to create new melodies that will sound good and as if they belong with the original melody. You will create new melodies by transforming the original melody using some simple mathematics operations on $f(x)$. For each transformation you will change the function, first by adding 2 to it, next by multiplying it by 2, then by both multiplying and adding, and last by multiplying by −1. For each transformation follow the process of making an in/out table, recording the notes on the staff with their corresponding numbers, and graphing the function on the Melodic-Contour Graph.

After each transformation your teacher will play the melody, and the class will discuss it. From the class discussion and your own observations, write down a name for each transformation and its key characteristics.

MELODIC VARIATIONS AND TRANSFORMATIONS

Number/Note Reference Key

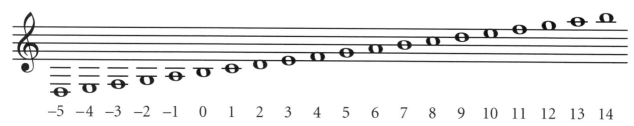

−5 −4 −3 −2 −1 0 1 2 3 4 5 6 7 8 9 10 11 12 13 14

Original Motif

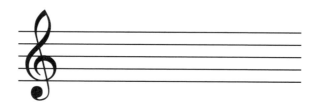

Melodic Variations

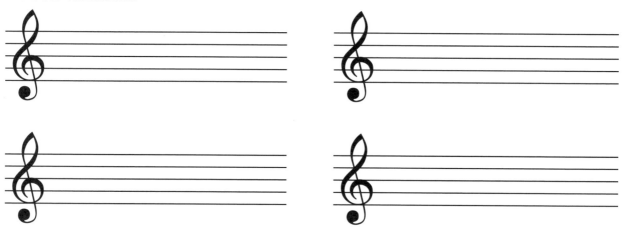

In/Out Table

x	$f(x)$	$f(x) + 2$			

MELODIC-CONTOUR GRAPH

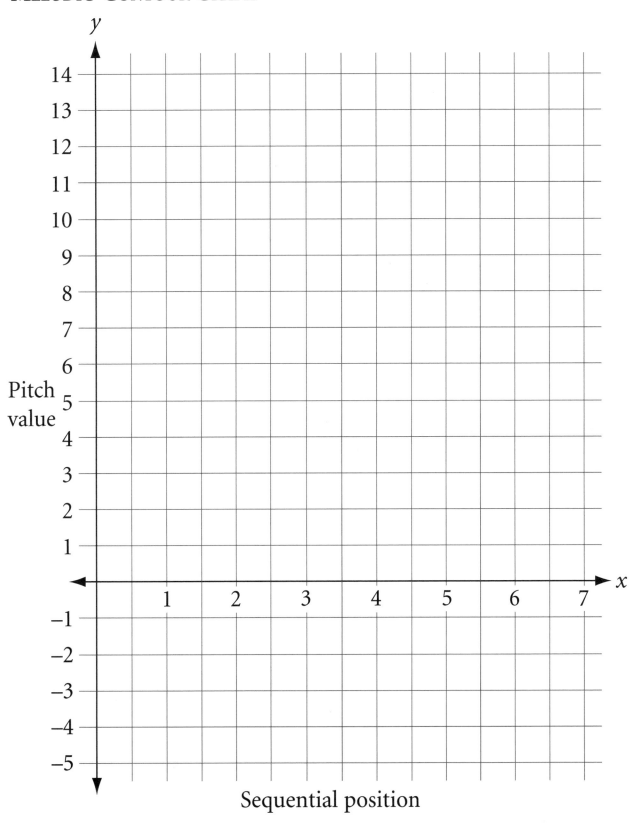

7

Functional Composer, Second Movement

The Relentless Composer

TRACKS
41–47

Functional Composer, Second Movement is a continuation of Functional Composer, First Movement, with some differences:

- This activity explores transformations resulting from operations on the input value of $y = f(x)$.

- The original motif is repeated to become a periodic function to allow students to see the transformations more clearly and to provide musical depth, both theoretically and aesthetically.

- The melodic variations are played simultaneously with the original motif on the music CD.

As in Functional Composer, First Movement, students calculate, graph, and listen to several transformations. Students make observations and conjectures about the transformations that can be generalized for all functions. The aesthetic appeal of the transformations is assessed and explained through analysis of the melodic-contour graphs. Horizontal translations are used to make adjustments to melodies that adhere to principles of melodic counterpoint. Students see that, as a result of the new transformations, the hypothetical composer now has more material to work with and more extensive tools available for extension projects.

Mathematics topics

Transformations of functions (translation, stretching, shrinking), periodic functions (phase shift, periods, amplitude, cycles), in/out tables, function notation, domain, range. *Prerequisites:* An understanding of function notation, use of in/out tables, and graphing on the Cartesian coordinate system.

Music topics

Melodic structure, ostinato patterns, melodic counterpoint, intervals. *Prerequisites:* Functional Composer, First Movement.

Use with the primary curriculum

- To introduce phase shift in trigonometric functions
  Use Functional Composer, Second Movement to help students understand the characteristics of periodic functions prior to textbook study of the graphs of trigonometric functions.

- To continue the study of graph transformations
  Transformations generated from operations on the input value of functions can be discovered or demonstrated as an introduction or follow-up application with Functional Composer, Second Movement.

Objectives

- To understand graph transformations in the *x*-direction
  Horizontal transformations (addition producing movement to the left) can be particularly counterintuitive for students. The process of calculating values and hearing the melodic representations reveals why this happens.

- To observe periodic functions
  Working with simple melodic shapes can lead to enhanced understanding of the elements of periodic functions.

Student handouts

- What to Do (resource page; one per group)
- In/Out Table and Conclusion (worksheet; one per pair)
- Melodic-Contour Graphs (worksheet; one per pair)

Materials

- CD tracks 41–47
- Overhead transparencies of student worksheets

Instructional time

30–50 minutes

Instructional format

Differences from Functional Composer, First Movement:

- Students have smaller graphs.

- Students do not represent melodies with notes on staffs—the melodies are treated purely as number sequences.

- The development of conclusions after each transformation is generated by students instead of by you.

Throughout the activity script, a line of questioning is suggested to provoke discussion among the students. Answers directly follow the questions and are in square brackets.

Student preparation

Functional Composer, First Movement should be completed a day or so before conducting the Second Movement. No additional background reading is necessary. If you decide to wait more than a week after doing the First Movement, you may need to review the ideas and the process of converting notes to numbers.

ACTIVITY SCRIPT

STEP 1   Introduction

The composer in Functional Composer, First Movement was not completely satisfied and needs more musical material. The relentless composer must now pursue another family of transformations.

Perhaps the question was raised in Functional Composer, First Movement as to what would happen if the input value, *x*, were operated on to transform the motif. If so, remind students of that question. Play track number 41 and ask students to follow along the graph to orient themselves to the motif. Unlike Functional Composer, First Movement, in this activity the motif is repeated.

Mathematics perspective: The motif is now a *periodic function*, that is, a function that repeats in cycles. Trigonometric functions are periodic functions. Anything that repeats is periodic, for example, tides, heartbeats, or seasons of the year. Each complete repetition of a pattern is called a *cycle*. The amount of time it takes for one cycle to repeat is called the *period*.

Ask students:

How many cycles of $f(x)$ do you see represented on the Melodic-
   Contour Graph? [2]
What is the period of $f(x)$? [6]
What is the range of $f(x)$? [integers $0 \le y \le 5$]
What is the domain of $f(x)$? [The interval of our study is integers
   $1 \le x \le 13$.]

Explain that the function could repeat as long as we like; but for our
study we will view just two cycles.

Music perspective: In music, a repeating melodic pattern is called an
ostinato pattern and is used often by composers, jazz improvisers, and
rock musicians.

STEP 2   The first transformation: $y = f(x + 2)$

Before they graph $f(x + 2)$, ask students to predict how adding 2 to the x-value
will change the graph. Suggest that they think about what they learned in
Functional Composer, First Movement. Here, instead of the melody becoming
higher, the starting note changes; the note that was third is now first. After some
discussion, have students complete the third column of the in/out table and
draw the graph on the first melodic-contour graph.

 Ask students:

Can you find a function value for $12 \le x \le 13$? [No.]
Why or why not? [The interval gives input values outside the
   specified domain of the original function.]

 Watch carefully as students insert the function values into the in/out
table. If they are not experienced with function notation, they may
not see that for $x = 1$, $f(x + 2) = f(3)$ and $f(3) = 2$. Students must
place 2 in the function column for $f(1)$.

After students have successfully filled in the in/out table and drawn the graph,
play track number 42. Have them observe how the transformation relates to the
original motif and record their observations about how adding to the input
value transforms the function.

 Do not feel that students must come to correct conclusions about the
transformation at this point. They may gain insight from later
examples. Let them discuss their observations with their partners, and
before the end of the activity a class discussion can verify accurate
conclusions.

STEP 3 Another transformation: $y = f(x - 3)$

Again, have students predict what the result of subtracting from the input value will be. Most will be able to make this prediction based on the previous transformation, $y = f(x + 2)$.

Have them calculate the in/out values and draw the graph. Then play track number 43.

After some discussion of the sound and the graph, have students record their observations and conjectures.

Ask students:

Which sounds better, $y = f(x + 2)$ or $y = f(x - 3)$? [Most will say that $y = f(x - 3)$ sounds better.]

Why? [It is more harmonious, less dissonant, and so on.]

Mathematics perspective: Have students look at the graph and notice the distance between the function values of $y = f(x + 2)$ and $y = f(x)$ at each x-value. When $x = 1, 2,$ or 3, this distance is 1. Notice that the distance between the function values of $y = f(x - 3)$ and $y = f(x)$ is never 1.

Music perspective: Melodic counterpoint is the interplay between two melodies played simultaneously. To write harmonious melodic counterpoint, composers need to consider how the distance between the notes of the two melodies changes as time progresses. For example, if the distance between the melodies is successively 1 note, the music will not sound harmonious. This is because successive small distances do not provide the tension and release necessary for the interplay of two melodies to be satisfying to our aesthetic sense. Later in this activity, students will use the melodic-contour graphs to adjust melodies to make them more harmonious.

Ask students:

What translation would give a function equivalent to the original motif? [$y = f(x + 6)$]

For any periodic function $f(x)$ where c is the period, $f(x) = f(x + c)$. This can be seen very clearly on the graph. Translation of a periodic function in the horizontal direction is called a *phase shift*.

Use two overhead transparencies of the original motif to demonstrate phase shift as horizontal translation. This is a great way to show that a phase shift of one period yields an equivalent function.

Before moving on, ask students to describe why subtracting from the input value translates the function in the positive *x*-direction while adding translates in a negative direction. Refer students to the in/out table for insight.

STEP 4 Another transformation: $y = f(2x)$

Have students calculate the in/out values and draw the graph. Then play track number 44.

After some discussion of the sound and the graph, have students record observations and conjectures.

Ask students:

Can you obtain function values when $7 \leq x \leq 13$? [No.]

Why or why not? [That interval for *x* yields input values outside the specified domain for the function.]

If you are studying trigonometric functions, point out that, because they are continuous, the graphs of $y = \sin(2x)$ and $y = \sin(x)$ look the same; the first just has half the period. The melodic function of $f(2x)$, however, being discrete, picks up only half the notes.

Direct students' attention to the relationship between the graph of the original motif and that of $y = f(2x)$. Recall the discussion of musical counterpoint and its connection to the graph of the melodies.

Ask students:

Considering the principles of melodic counterpoint discussed earlier, can you think of a way to make $y = f(2x)$ sound better when played with the original motif? [At several points, the two melodies have distances of 1 between them. If $y = f(2x)$ were translated in either direction, there would be fewer places at which the melodic contours are just 1 unit apart and thus the two motifs might be more harmonious.]

What (horizontal) translation of $y = f(2x)$ would result in no distances of 1 between the original function and the transformed function? [Translating $y = f(2x)$ 1 unit to the left or 2 units to the right are the only choices.]

Does translating $y = f(2x)$ to the left by 1 unit or the right by 2 units create different graphs? [No.]

Why or why not? [The period of $y = f(2x)$ is 3, and these two translations are the same shape translated 3 units from each other. Notice that multiplying the input value by 2 has divided the period of the original function by 2.]

How would these translations be written? [$y = f(2(x + 1))$ and $y = f(2(x - 2))$]

Students may suggest that the translations would be $y = f(2x + 1)$ and $y = f(2x - 2)$. Point out that these are actually different functions from the correct responses. Have students calculate some values to verify this. You can review order of operations and composite functions while pointing out that a translation occurs only when a value is added to or subtracted from the input value before any other operation is performed, hence the need for parentheses around $x + 1$ and $x - 2$.

CD track number 45 is the melody created by $y = f(2(x - 2))$. Play this track. The differences in the harmonious qualities between $y = f(2x)$ and $y = f(2(x - 2))$ may be difficult for students to hear. Play it several times.

Ask students:

Does this variation sound better (more harmonious) than $y = f(2x)$? [Yes.]

Why or why not? [It has been translated 2 units to the right, so the distances between function values in the two motifs are no longer 1.]

STEP 5 Another transformation: $y = f\left(\frac{1}{2}x\right)$

Have students calculate the in/out values. There will be points for $f(x)$ only if x is an even number. Have them draw the graph on the last melodic-contour graph. Then play track number 46.

After some discussion of the sound and the graph, have students record their observations and conjectures.

STEP 6 A composition and analysis

As with Functional Composer, First Movement, this activity concludes with a musical composition that uses several transformations. This track combines transformations from Functional Composer, First Movement and Functional Composer, Second Movement. After summarizing the transformations graphed in the activity, play the track several times, if necessary, and ask students to identify the transformations they hear and identify where they occur.

Notice that horizontal melodic translations can be difficult to pick out by ear, since they are shifted only in time. The horizontal stretching and shrinking will be easier to identify, as of course will be the vertical transformations. Some minor melodic material that is not a transformation of the original is occasionally used in the composition to connect ideas. The composition on the CD is in two distinct sections with some rhythmic shape added to create interest. In the first section, the lower melody is created from the function $y = -f\left(\frac{1}{2}(x+1)\right) - 2$ alternating with $y = f\left(\frac{1}{2}(x+1)\right)$. The higher melody is simply $y = f(x)$ repeated. In the second section, the higher melody uses a series of vertical translations of $y = f(2x)$. The lower part of this section uses $y = -\left(\frac{1}{2}(x+1)\right) - 7$ and $y = -f\left(\frac{1}{2}(x+1)\right) - 2$.

FOLLOW-UP ACTIVITIES

Many of the follow-up activities for Functional Composer, First Movement will also apply to Functional Composer, Second Movement. Some specific suggestions for Functional Composer, Second Movement are listed here.

Textbook assignments

Following the activity with textbook assignments can help students connect the ideas directly to a pure mathematical context. Trigonometric graphing problems or function transformation problems of any kind connect well to this activity.

Project

Use all the transformations studied to make a piece of music. Represent the music graphically, and have it played in class either by a computer or by several students. Give an analysis of how mathematics was used to create harmony, interest, variety, and structure. A project such as this could be combined with a project in a music class.

ANSWERS

In/Out Table and Conclusion

In/Out Table					
x	$f(x)$	$f(x+2)$	$f(x-3)$	$f(2x)$	$f(\frac{1}{2}x)$
1	1	2	—	5	—
2	5	4	—	4	1
3	2	3	—	0	—
4	4	0	1	5	5
5	3	1	5	4	—
6	0	5	2	0	2
7	1	2	4	—	—
8	5	4	3	—	4
9	2	3	0	—	—
10	4	0	1	—	3
11	3	1	5	—	—
12	0	—	2	—	0
13	1	—	4	—	—

Conclusion		
Transformation	**Observation**	**Conjecture**
$f(x+2)$	**Graph** translates to the left by 2 units. **Melody** sounds the same, has the same notes, but occurs sooner in time.	Adding a value, h, to the input value of a function translates the graph of the function h units to the left.
$f(x-3)$	**Graph** translates to the right by 3 units. **Melody** sounds the same, has the same notes, but occurs later in time.	Subtracting a value, h, from the input value of a function translates the graph of the function h units to the right.
$f(2x)$	**Graph** shrinks in the horizontal direction. **Melody** has some notes left out, becomes simpler.	Multiplying the input value of a function by a value, c, shrinks the function in the horizontal direction by a factor of $\frac{1}{c}$.
$f(\frac{1}{2}x)$	**Graph** stretches in the horizontal direction. **Melody** sounds slower, it has all the original notes, but they are spread out over time.	Dividing the input value of a function by a value, c, stretches the function in the horizontal direction by a factor of c.

TEACHER NOTES

Melodic-Contour Graphs

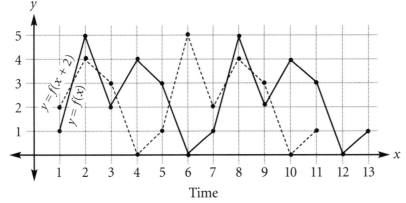

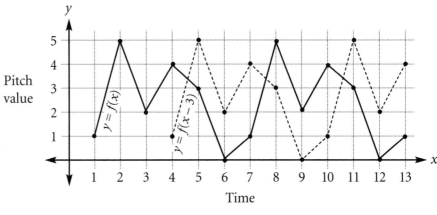

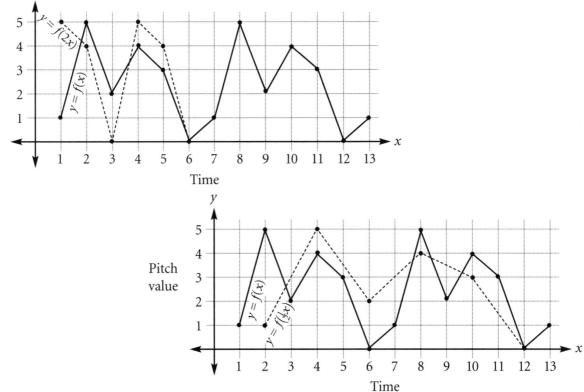

WHAT TO DO

You helped a composer with writer's block by using transformations of a function to generate melodies for a composition. Now you are going to continue that process by transforming the functions in a different way: operating on the input value, x. As in Functional Composer, First Movement, you will transform a function based on a musical motif to create melodic variations.

Listen to the original motif.

The original motif has been slightly altered from what was used in the First Movement. It is now repeated, and the first note has been changed. The notes (which are the same as the function values) are 1, 5, 2, 4, 3, 0, and this sequence repeats once. Your teacher will play the melody.

Create melodic variations/function transformations of the original motif.

The transformations being explored in today's activity operate on the input value, x, of the function $y = f(x)$. For the first melodic variation, $y = f(x + 2)$, you will add 2 to x (when x is 1, $f(x + 2)$ will be $f(3)$) and find that $f(3)$ is 2. Therefore, for $x = 1$, $f(x + 2) = 2$. Fill out the rest of the corresponding column in the in/out table by adding 2 to x and finding the value of $f(x)$ for that value. When the in/out table is complete, graph the transformation on the first melodic-contour graph. Your teacher will play the CD track for that transformation. Write your observations and conclusions.

What do you see?

What do you hear?

What, in general, will be the effect of adding a number to the x-value?

To find other new melodic contours, you will use three more transformations: subtracting 3 from x, multiplying x by 2, and multiplying x by $\frac{1}{2}$.

For each transformation, fill in the in/out table, graph the melodic contour, listen to the CD track of that new melody played with the original motif, and write down your observations and conjectures.

IN/OUT TABLE AND CONCLUSION

In/Out Table					
x	$f(x)$	$f(x+2)$	$f(x-3)$	$f(2x)$	$f(\frac{1}{2}x)$
1					
2					
3					
4					
5					
6					
7					
8					
9					
10					
11					
12					
13					

Conclusion		
Transformation	Observation	Conjecture
$f(x+2)$	Graph Melody	
$f(x-3)$	Graph Melody	
$f(2x)$	Graph Melody	
$f(\frac{1}{2}x)$	Graph Melody	

MELODIC-CONTOUR GRAPHS

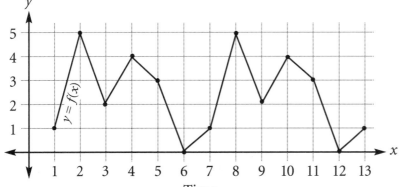

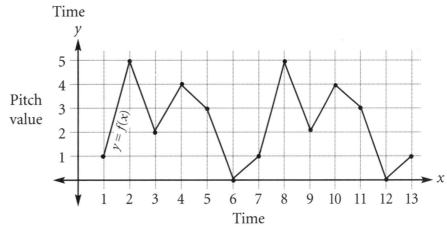

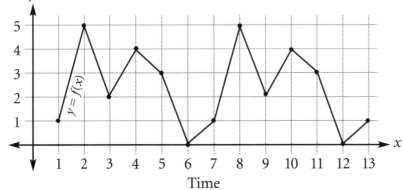

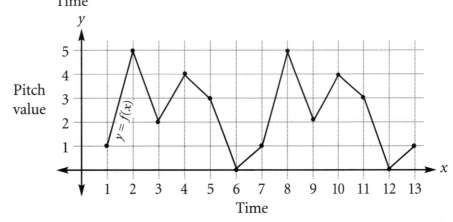

Name That Function

Determining Function Transformations by Listening

TRACKS 48–72

Name That Function is a collection of six activities that use function transformations similar to those in Functional Composer, First and Second Movements. The activities follow the format of a guessing game as in the old television show *Name That Tune*, where contestants competed to name a tune from the first few notes. Here students apply their knowledge of functions and mathematical relationships to make strategic, informed guesses about the identity of a mystery function. In each game, a melodic line is played, followed by a mathematical transformation of that line. The initial goal is for students to determine the equation of the mystery transformation exclusively by ear. To assist this process, the notes of the original and transformed melodies are assigned number values, treated as functions, and

compared. Ultimately students use note numbers (with in/out tables and graphs if necessary) to determine the exact equations of the transformations. Each game involves four transformations of a single motif.

The games are versatile; they can be used individually as short warm-up exercises or used together in an entire class session. The process can also be reversed, with students attempting to sing the transformations from their equation representations.

Unlike Functional Composer, this activity does not require graphing. Name That Function emphasizes the melodic aspect and forces students to resort to a mathematical model (a graph or calculation) of their own devising as a tool for determining the equations.

Mathematics topics

Function notation, transformations of functions (stretching, shrinking, reflecting, translating), calculating function values, problem solving.
Prerequisites: Functional Composer or experience with function notation.

Music topics

Melodic dictation/ear training, interval recognition, musical notation.
Prerequisites: Students with no musical background will benefit from having done Functional Composer.

Use with the primary curriculum

- As a warm-up exercise before any class session
  Each game presented in Name That Function can be conducted in as little as 10 minutes as a warm-up for any class session.

- For enrichment and review of functions and transformations
  Use Name That Function during the study of functions of any type (trigonometric, logarithmic, quadratic, linear, conic sections) to review transformation principles.

Objectives

- To create a fun atmosphere for any class session
  Used as a warm-up exercise, Name That Function can capture the attention of students and establish an atmosphere that can enhance any day's instruction.

- To reinforce the understanding of functions and transformations
  The nonroutine context for function transformations used in Name That Function requires students to think through the ideas in new ways, enhancing their understanding and retention.

Student handouts

- Hints and Strategies (resource page; one per group)
- Answers and Calculations (worksheet; one per student per game)

Materials

- CD tracks 48–72
- Overhead transparencies of Melodic Transcriptions of Function Transformations
- Graph paper (optional)

Instructional time

10–50 minutes

Instructional format

The six activities, referred to as games, follow the steps outlined in the activity script. Divide your class into groups of two to four students; do your best to distribute musically experienced students equally among the groups.

Name That Function is highly versatile. Be creative and adaptable. With any variation, you can reverse the process: Play the original motif, give students the equation of the function transformation, and have a brave soul sing the transformation. You can do this initially as a qualitative exercise using mental approximations of what the pitches should be, followed by students calculating the values and singing from melodic transcriptions. Finally, check the vocal performance against the CD track.

Transformations in Games 1 through 4 require operating on the output value of a function, as in Functional Composer, First Movement.

Game 5 uses transformations on the input value, as in Functional Composer, Second Movement.

Game 6 combines transformations on both the input and the output values. If you have completed Functional Composer:

- Use Name That Function as a 10-minute warm-up exercise with no handouts. A game is a great way to open any class session and revisit ideas throughout the course. It does not need to lead into a lesson on functions. By revisiting applications of functions many times throughout a course, your students will deepen their understanding and retain more of what they have learned. Students can use scratch paper instead of handouts when they have become familiar with the format.

- Use Name That Function with handouts and transcriptions of the melodies. A game of Name That Function can take up to 30 minutes if students are given a few hints along with the worksheets and they use note numbers to make musical transcriptions of functions. Transcriptions are needed for the more difficult games—and indeed for all games with students who are less mathematically or musically advanced.

If you have not completed Functional Composer:

- First conduct a condensed version of Functional Composer. Students need to understand how to convert notes to numbers, graph them, and treat them as a mathematical function. Once they understand the process, they will be ready to complete a game of Name That Function using the handouts.

TEACHER NOTES

Student preparation

There is no preliminary reading for these games, but if you have not done Functional Composer, have students read its preliminary reading, A Composer's Problem.

ACTIVITY SCRIPT

The script describes running a 30-minute game using the worksheets. The procedure for all games is the same, though specifics vary. Once you understand the procedure, use the coaching notes in the answer key for specifics about a particular game.

Unless a transformation is simple or a student is very musically skilled, many students will have difficulty determining the transformations exclusively by ear from the CD track. Each successive step of the script reveals more information and tools to assist them. Apply these steps strategically to maximize the discovery for all students in your class. Observe students and provide whatever amount of time they need. For some transformations or some classes, not all of the steps will be necessary. Apply the given sequence of steps to each transformation as appropriate.

Discovery of the mystery function can be broken down into two stages: determining the types of transformations and then determining the exact mathematical equation. Identifying the transformation type (a translation, a stretching or shrinking, or a reflection) will often be possible by ear, while determining the exact equation will require deeper analysis.

STEP 1   Introduce the concept of the game

Give students an overview of the game and review the concepts of Functional Composer, especially how the function transformations of shrinking, stretching, translating, and reflecting are expressed in a mathematical equation in function form.

Note that the student resource page, Hints and Strategies, provides key questions rather than direct information. The questions will direct students' inquiry and discussion rather than hand them information. You will need to supply the answers to these questions when students are unable to, because students will need the information in order to proceed with the activity.

STEP 2   Play the original motif and first transformation

The CD track for each transformation includes the original motif played before the transformation for comparison. Play the track for the first transformation and wait for student response. Do not show students the melodic transcription yet. To help students determine the type of transformation, ask them to listen to the relative distances between the notes of the transformation compared to the distances in the original motif. If the distances are the same, the transformation is a translation (adding or subtracting from the function). If they are different, the transformation is a stretching (multiplying the function value). Students should also pay attention to the overall pitch. If it is higher, the melody has been translated up; if it is lower, the melody has been translated down. If the melodic movement is in a direction opposite that of the original, the transformation is a reflection.

Many students will not be able to hear the pattern from the CD track alone. The next step will help them, but don't rush to move on. Repeat the track as many times as necessary. All students should make a thoughtful guess at this point. They will test their guesses by using note numbers when they complete the activity.

To ensure that all students have the opportunity to make their own connections, ask them to remain quiet about their discoveries until all students have made informed guesses. They will be eager to share their discoveries, but ask them instead to make a gesture indicating that they have reached a conclusion. It is not unusual for a few students to be especially good at this game and discover the equations quickly; don't let them dilute the discovery experience for the rest of the class.

As noted in Functional Composer, translations of the melody here are not exactly accurate mathematical translations, since the distances between notes are not consistent in a major scale. These translations are different modes. An exact translation would require the use of sharps and flats. An approximation is made in order to streamline the musical content of the activity and to allow students to focus on the mathematics.

It may be helpful for some students to listen for common notes between the original and the transformation and to determine where they occur. Cross-referencing other notes can help students discover how much a melody was translated, but recognizing specific musical intervals requires a developed ear. Referencing common notes will be easier for most students when they have the musical transcriptions and the numbers for the notes. Track 48 plays the notes of the scale used in the number/note reference key. It can be used to help students make more accurate estimations by ear of the number of scale steps between notes and to determine the number value for at least one note for reference.

TEACHER NOTES

STEP 3　Reveal the melodic transcriptions

At this point you might reveal only the original motif to allow students another chance to determine the note values by ear and find the equation of the transformation.

Display the melodic transcriptions of the original motif and translation and have students copy the melodic lines to the staff on their worksheets. The note values have purposely been left off the transparency masters to allow students to be creative in how they use the melodies to determine the equation.

Direct students' attention to the geometry of the notes on the staff and to how the shape depicts relative distances (musical intervals) visually, like a graph. For students to see this effectively, they will need to notate the transformation on the same staff as the original. Students' knowledge of the graphs of function transformations can aid them in determining types of transformations directly from looking at the melodic transcription, without using numbers.

STEP 4　Label pitch values on the melodies

In many cases students will need to assign pitch values to the melodies as a first step toward establishing the equation for the transformation. Don't be too directive. Instead, let students realize this need themselves.

Once students have found the numbers for the notes on the worksheet and have determined the type of transformation, the process of determining the equation becomes purely mathematical. Urge students to make use of common notes, along with their knowledge of the translation types, to apply thoughtful strategies. For example, if a transformation's shape indicates that it is a reflection but the numbers are not exclusively the negatives of each other, then it was also translated. The amount of translation can be determined by comparing a nontranslated reflection with the translated reflection. When stretching is added, the process can become more complicated, and students may resort to a wide variety of strategies, including guess-and-check. Placing the numbers for each melody in an in/out table and graphing them to see the relationship clearly, as in Functional Composer, can be a valuable tool.

STEP 5　Explain the solution

Direct students to explain their method of solution in words and to show any dead-end guesses and checks that helped them arrive at their equation.

STEP 6   Repeat the process for the other three transformations

Follow the process in Steps 2–5 for each of the other three transformations of the function. As before, give students an opportunity to guess the function from what they hear, then display the melodic transcription. Students can label the pitch values and determine the equation of each transformation.

Use CD tracks 53–56 for Game 2; CD tracks 57–60 for Game 3; CD tracks 61–64 for Game 4; CD tracks 65–68 for Game 5; and CD tracks 69–72 for Game 6.

FOLLOW-UP ACTIVITIES

Functional Composer, Second Movement

After students have done Games 1 through 4, Functional Composer, Second Movement is an ideal follow-up activity and will prepare students for Games 5 and 6.

Textbook assignments

Follow Name That Function with textbook assignments that apply graph transformations in any context—trigonometric functions, logarithmic functions, conic sections, or geometric mapping—to reinforce the concepts.

Project

Students can invent games and write music using more complex functions, applying transformations to more rhythmically complex melodies to find pleasing combinations. Follow up with a class presentation of how the transformations work on the melodies and why some are more pleasing than others.

ANSWERS

Vertical transformations—operating on the output value

Games 1 through 4 use transformations on the output value. On the CD tracks, the original motif and the transformation are played separately, one after the other. In vertical transformations, time is not relevant. These transformations will be compared according to their pitch relationships.

Game 1

Transformation	CD track	Transformation equation	What to listen for in the transformation
1	49	$y = f(x) + 2$	Relative distances and the direction are the same, but overall it is higher in pitch. The first note is two notes higher than the first note of the original. The second note is the same as the third note of the original.
2	50	$y = 2f(x)$	Relative distances are greater, but the direction is maintained—the original has been stretched. The second note is the same as the third note of the original.
3	51	$y = -f(x)$	Relative distances are the same, but the direction is opposite. The fifth note (pitch value 0) is the same as the fifth note of the original.
4	52	$y = 2f(x) + 2$	Relative distances are greater, direction is maintained, and it is higher in pitch overall. The fifth note is two pitches higher than the fifth note of the original, which has a pitch value of 0.

Game 2

Transformation	CD track	Transformation equation	What to listen for in the transformation
1	53	$y = f(x) - 4$	Relative distances and the direction are the same, but it is shifted lower in pitch. The fourth note is the same as the third note of the original.
2	54	$y = 2f(x) - 3$	Relative distances are larger, the direction is the same, and it is shifted lower in pitch. The last notes are the same.
3	55	$y = -f(x) + 10$	Relative distances are the same, the direction is opposite, and it is shifted higher in pitch overall. The second note is the same as the second note of the original.
4	56	$y = -2f(x) + 12$	Relative distances are larger, the direction is opposite, and the pitches are shifted higher. The last note is the same as the fourth note of the original.

Game 3

Transformation	CD track	Transformation equation	What to listen for in the transformation
1	57	$y = -f(x)$	Relative distances are the same, and the direction is opposite. The fourth note is the same as the fourth note of the original.
2	58	$y = f(x) + 6$	Relative distances and the direction are the same. The pitches are shifted higher. The fifth note is the same as the seventh note of the original.
3	59	$y = 3f(x)$	Relative distances are much larger, and the direction is the same. The third note is the same as the first note of the original. The fourth note is the same as the fourth note of the original.
4	60	$y = -2f(x) + 8$	Relative distances are larger, and the direction is opposite. The fourth note is not the same as the fourth note of the original.

Game 4

Transformation	CD track	Transformation equation	What to listen for in the transformation
1	61	$y = -f(x) + 6$	Relative distances are the same, but the direction is opposite. The starting note is the same as that of the original.
2	62	$y = 2f(x) - 3$	Relative distances are greater, and the direction is the same. The starting note is the same as that of the original.
3	63	$y = 4f(x) - 6$	Relative distances are very much larger, and the direction is the same. The third, sixth, and seventh notes are the same as those of the original.
4	64	$y = -2f(x) + 9$	Relative distances are larger, and the direction is opposite. The first two notes are the same as the original.

TEACHER NOTES

Horizontal transformations—operating on the input value

Do not announce to students that the transformations in Games 5 and 6 are different from those in Games 1 through 4. Let them discover the difference on their own.

Since the horizontal axis represents time, the only meaningful way to compare these transformations is to play them in relation to each other in time. Consequently the CD tracks for input-value transformations are presented with the melodies overlapping. For the functions in Games 1 through 4, the input value is the sequential number of the note. In Games 5 and 6, this sequential number corresponds to the beat—in this case, one quarter note. The original motifs are repeated as a periodic function to enable the horizontal translations to be seen more clearly, as in Functional Composer, Second Movement.

Game 5

This game uses the motif from Game 2.

Transformation	CD track	Transformation equation	What to listen for in the transformation
1	65	$y = f\left(\frac{1}{2}x\right)$	Relative distances are the same, but it is stretched out over time. Notes are not played on odd-numbered beats. This implies that input values not evenly divisible by 2 are unusable, since the domain of the function is integers only. Thus the transformation must be dividing the input values by 2.
2	66	$y = f(2x)$	The melody is shortened—compressed over time—and alternate notes of the original are missing in the translation.
3	67	$y = f(x + 1)$	The melody sounds the same in all respects except that it begins with the second note of the original; it is shifted earlier in time by one beat.
4	68	$y = f\left(\frac{1}{2}(x + 1)\right)$	This melody sounds the same as $f\left(\frac{1}{2}x\right)$, but it starts one beat sooner; it is shifted one beat earlier in time.

Combining vertical and horizontal transformations

Game 6

This game uses a new motif.

Transformation	CD track	Transformation equation	What to listen for in the transformation
1	69	$y = f\left(\frac{1}{2}x\right) - 4$	Relative distances are the same, but it is stretched out over time. The notes are lower in pitch overall.
2	70	$y = f(2x) + 5$	Relative distance is difficult to compare, since the melody has been compressed in time with some notes left out. The notes are higher in pitch overall.
3	71	$y = 2f\left(\frac{1}{2}(x + 1)\right)$	Relative distances are larger, and the melody has been spread out over time. This melody begins one beat sooner than $y = f\left(\frac{1}{2}x\right)$.
4	72	$y = 2f(2(x - 2)) - 3$	Relative distances are greater and the melody has been compressed in time with some notes left out. The melody occurs later in time than the original, and the pitches have been stretched overall.

MELODIC TRANSCRIPTIONS OF FUNCTION TRANSFORMATIONS

Game 1

Original motif: $y = f(x)$

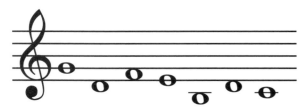

Transformations:

1.

2.

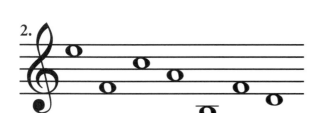

3.

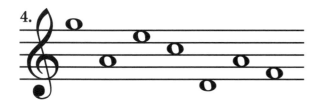

4.

Game 2

Original motif: $y = f(x)$

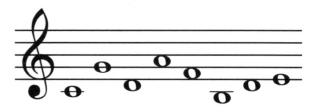

Transformations:

1.

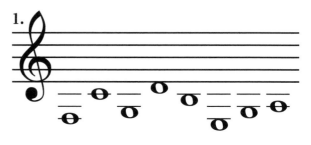

2.

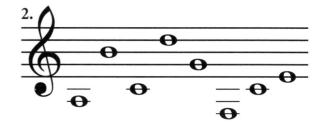

3.

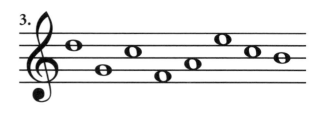

4.

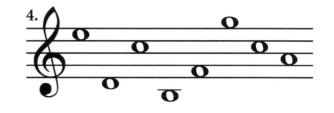

MELODIC TRANSCRIPTIONS OF FUNCTION TRANSFORMATIONS

Game 3

Original motif: $y = f(x)$

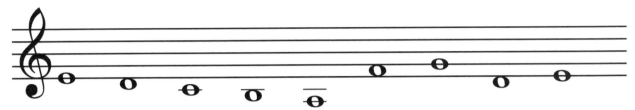

Transformations:

1.

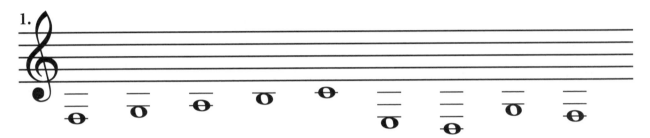

2.

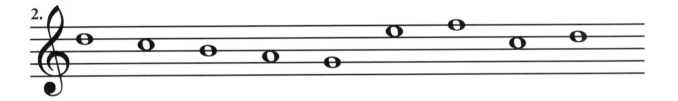

3.

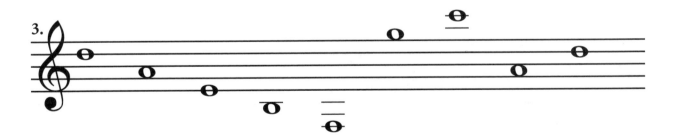

4.

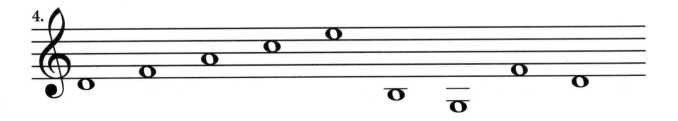

MELODIC TRANSCRIPTIONS OF FUNCTION TRANSFORMATIONS

Game 4

Original motif: $y = f(x)$

Transformations:

1.

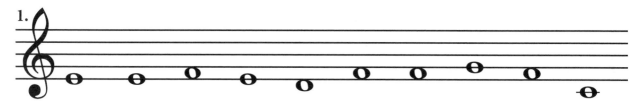

2.

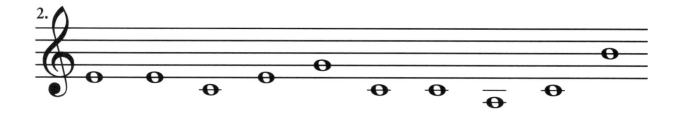

3.

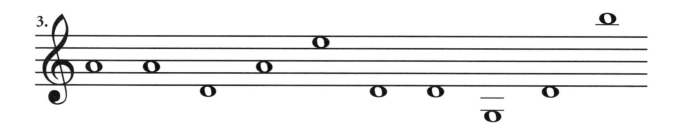

4.

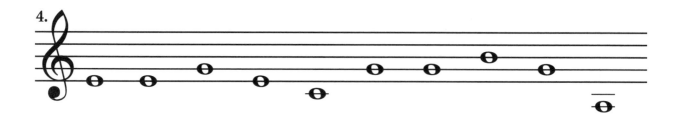

MELODIC TRANSCRIPTIONS OF FUNCTION TRANSFORMATIONS

Game 5

Original motif: $y = f(x)$

Transformations:

1.

2.

3.

4.

MELODIC TRANSCRIPTIONS OF FUNCTION TRANSFORMATIONS

Game 6

Original motif: $y = f(x)$

Transformations:

1.

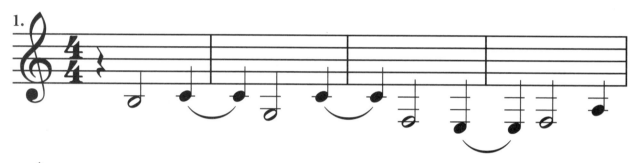

2.

3.

4.

HINTS AND STRATEGIES

Listen to the motif and transformation played from the CD.

Determine how the motif has been transformed.

Listen to how the notes compare to the original.

- Do they move in the same direction?

- Do they move the same distances?

- Are they shifted higher or lower overall?

Describe the types of graph transformations.

- What is a translation?
- What is a reflection?
- What is a stretching?
- What is a shrinking?

With your group, talk about how these transformations would sound when applied to a melody.

Determine the equation for the transformation by listening to the CD several times.

Listen to the major scale for reference if this will help you.

- How do the first notes of each melody compare?

- How many scale steps higher or lower are they?

- Are there common notes between the melodies?

Compare specific notes in their melodic sequences. For example, if a stretching transformation creates the same note as a translation, you know that some number times that note equals some number plus that note.

Determine the equation for the transformation by using numbers.

Use the tools on your Answers and Calculations worksheet to assign numbers to the melodies, check the estimations you made by ear, and find the exact equation for the transformation.

     Use in/out tables, graphs, or another method.

Explain how you determined the equation.

For each transformation of the original motif, show all your work and explain in words the strategy you used.

ANSWERS AND CALCULATIONS: GAME _____

Number/Note Reference Key

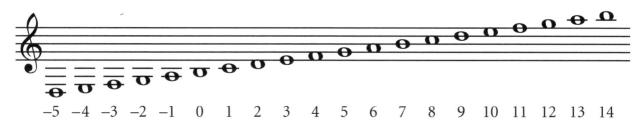

−5 −4 −3 −2 −1 0 1 2 3 4 5 6 7 8 9 10 11 12 13 14

Original motif: $y = f(x)$

Transformation 1: $y =$ _____
Explain how you determined this function.

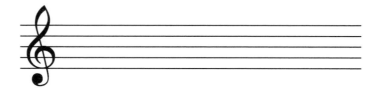

Transformation 2: $y =$ _____
Explain how you determined this function.

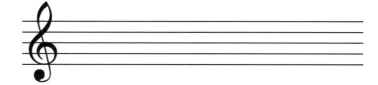

Transformation 3: $y =$ _____
Explain how you determined this function.

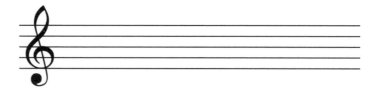

Transformation 4: $y =$ _____
Explain how you determined this function.

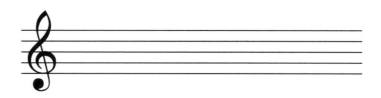

9

Inside Out

Hearing Pictures As Music Through Polar Coordinates

By using polar coordinates as an interface, Inside Out allows students to make music from graphic images. Students trace the outline of an image onto a polar coordinate system and determine the intersection points of the image and the graph at degrees 0, 10, 20, 30, and so on. They convert the polar distance of each point (r-value) to a corresponding musical pitch using the conversion key from Functional Composer. The notes correspond to the polar distances, while their order corresponds to the successive angle measurements. The result is a melodic line that embodies the shape of the graphic image. The melodies are performed by students; by listening to the melodies and viewing the original graphic images, the class matches each melody to the visual image that created it. The matching process becomes an exercise in which students must visualize how a ray with a constant angular velocity changes length over time. Students also create melodic lines from geometric figures and polar curves.

Since the students themselves create the melodies, at least one instrument and an instrumentalist who can read music will be needed for playing the unique melodies. You can illustrate the concept using the CD track for the fish example.

Mathematics topics

Polar coordinates and polar graphing, scaling of axes, angles, angle measurement, mathematics as an interface between two media. *Prerequisites: Some experience with coordinate graphing, polar or rectangular.*

Music topics

Musical notation, staffs, melodic contour, composition, imagery in music, instrumental performance. *Prerequisites:* At least one person who can read music to play the melodies.

Use with the primary curriculum

- To introduce polar coordinates in geometry and algebra 1
  While polar coordinates are not standard topics for geometry and algebra 1, introducing them with Inside Out can enhance students' ability to absorb the topic more fully in subsequent courses.

- To reinforce polar coordinate understanding in algebra 2
  When polar coordinates are studied as part of the primary curriculum, Inside Out can provide an intriguing application while reinforcing concepts.

- To show how mathematics can connect to the arts
  Used in any mathematics, music, or art class, Inside Out can provide a compelling example of how mathematics, images, and music are related.

Objectives

- To introduce the polar coordinate system
  Inside Out provides a fun and engaging context in which to learn about polar coordinates for the first time.

- To deepen understanding of the polar coordinate system
  The task of matching visual images to melodies created using a polar coordinate interface stretches students' ability to visualize the connection between polar curves and the changing angles and ray lengths that create them.

- To let students experience mathematics as a unifying tool and to stimulate their imagination
  Seeing how mathematics can be used as an interface between different media can motivate, inspire, and expand students' awareness of and appreciation for mathematics.

Student handouts

- From Sight to Sound (reading; one per student)
- Information Center (resource page; one per pair)
- Polar Coordinate/Musical Note Interface: 0°–180° (worksheet; one per pair)
- Polar Coordinate/Musical Note Interface: 190°–360° (worksheet; one per pair)
- Polar Coordinate Graph (worksheet; one per pair)
- Rectangular Coordinate/Musical Note Interface (worksheet for optional activity extension; one per pair)

Materials

- CD tracks 73–76
- Overhead transparencies of student worksheets
- Overhead transparency of Polar Coordinate Graph: Fish example
- Overhead transparency of Rectangular Coordinate/Musical Note Interface: Mountain example (optional activity extension)
- Musical instrument
- Assorted graphic material from magazines, catalogs, personal photographs, and so on (from you or students)

Instructional time

20–150 minutes

Instructional format

Inside Out follows a relatively open, project-style format. After the process and goal are made clear, students work in groups at their own pace to develop melodies from the graphic images. At the end, students come together as a whole class for performances, melody-image matching, and discussion.

Group students in threes so that the number of melodies to play will not be unwieldy. Most students will want their melody performed, and listening to and matching more than about ten melodies to corresponding images may become boring and tedious for some classes. Adding accompaniment and rhythm as described in Step 6 of the activity script will make the melodies more interesting and help keep students' attention for the performances of all the class's melodies.

You will need to identify musicians in the class who can perform the melodies and assist in providing rhythm and other embellishments. Designate these students as "musical experts" and distribute them as evenly as possible among the work groups.

Inside Out can take from one 20-minute session to three 50-minute class sessions depending on the approach used. Feel free to create your own version of the activity. Three possibilities are listed here:

- Short 20-minute presentation
  If you do not wish to invest a lot of time but want to expose your students to the ideas of the activity, present just the fish example.

- One or two class sessions
  You can provide graphic images that are compatible with Inside Out and have students spend one day learning the system, tracing images, creating melodies, and performing some of them. You will need at least part of another day for students to perform the balance of the melodies and discuss the outcomes.

- Two or three class sessions
  If students bring in their own graphic images (such as photographs, graphic designs, or logos), you will first need to introduce them to the graph of the fish example so they can see the size and style of image that will work for the activity. Tracing images, creating melodies, embellishing them with variations, performing them, and discussing results will fill two class sessions beyond the introduction day. Using their own images inspires students but requires a larger time investment.

Consider collaborating with an art or music teacher to make Inside Out a joint assignment. This will increase the importance of the project and also model how artists and mathematicians might work together to develop a common product.

Student preparation

Have students read From Sight to Sound. If students are providing the graphic images, you will have to do Steps 1 and 2 of the activity script several days prior to the rest of the activity.

ACTIVITY SCRIPT

Because students are working in groups and at their own pace, the resource page, Information Center, is longer and denser than the resource pages in other *Functional Melodies* activities. It includes most of the essential information students need in order to complete the activity. Find your own balance between how much you direct the steps of the activity script and how much you let students work though the activity on their own, guided by the resource page. Regardless of how much the students make use of the resource page, it can be a useful outline and reference for you.

An example

The example of a fish graphic and its melodic counterpart (CD track 73) are included primarily as a reference for you to see and hear how the activity plays out. Unless you plan to conduct a short version of the activity, do not demonstrate this example in its entirety with the students before the activity. Seeing and hearing a final product will dilute the element of mystery and discovery. You could play it as another example after the activity is completed or show students the fish graphic before the activity as an example of the kind of image that translates well.

The example uses only 180 degrees of the image, and some musical embellishment is added in the CD track. The bass line is created from the fish image and the higher string melody is a transformation of the fish melody using the function $y = -f(x) + 11$.

STEP 1   Discussing the reading and defining the activity

After students read From Sight to Sound, start a discussion with questions.

> *Ask students:*
> What points did the reading make?
> What other kinds of interfaces do you know of?
> What do you think about using mathematics to create music?
> Are there ways that art conveys things that can't be understood using
>   mathematics?

Give the students an overview of the activity by reading through the headings of the resource page, Information Center. They will need to have a good sense of what they are trying to accomplish in order to choose appropriate images.

STEP 2   Choosing a graphic image

If students bring in their own images, they need to be aware of some limitations. Graphics that cannot be reduced to a defined silhouette will not translate well to distinct coordinate values. Images that are essentially round with little variation will produce potentially boring melodies, possibly one repeated note. Student awareness of this will be spotty until after they have done the activity; having some images that produce boring melodies can be instructive as a point of discussion. A good way to equip the class to choose appropriate images is to provide examples of images that will not work well. On the other hand, the fish example is a good image because its contour is distinct, it has some variation in its form, and it fits comfortably on the coordinate graph.

 Depending on students' understanding of the polar coordinate system and how it is used in this activity, their criteria for choosing images will vary. Some may focus on an image that they think will sound

TEACHER NOTES

good. Others will focus on images that they like for one reason or another or that have personal significance. Whatever their criteria, have them justify their choice thoughtfully.

Since you need to limit the number of images used if all the melodies are to be performed, each student group will need to agree on one image to use.

STEP 3   Tracing the images to the polar coordinate graph

If students find it difficult to see through the paper well enough to trace their image to the polar graph, they can cut out the image and use it as a stencil. Generally the center of the image should be near the pole of the graph, but students should feel free to experiment.

 Ask students how the melody would change if the position of the image on the graph were shifted. This will be difficult for them to imagine before they do the activity, but it is valuable to float the question at this point and come back to it after the melodies have been played.

STEP 4   Determining the polar coordinates for the image

Each point at which the image intersects the rays emanating from the pole needs to be entered on the coordinate/note interface chart. The angles are already printed, so students simply need to count off the distance on each ray. This requires that students understand how to locate and represent points with polar coordinates and how to scale the polar graph for their image.

These and the other tasks are described on the resource page. Let students work through them using the resource page. Intervene with assistance or direct instruction only when students need help.

Scaling the graph will depend on the size of the image. Generally, if an image reaches to the outer edges of the graph, a scale of one note per unit distance from the pole will be appropriate. If an image is relatively small, a scaling of two notes per unit distance will make use of the full range of notes.

 Explore with students the idea of scaling. Ask them how scaling will affect the resulting melody in general, and then ask what scaling would be appropriate for their image. Many will find it hard to make a judgment about this without having experienced the activity or polar coordinates. A scale of two or more notes per graph unit will create a melody with high contrast, and for many images the range might extend beyond the available notes. A scale of one half of a note or less per graph unit will create a melody with little contrast and limited range.

The Polar Coordinate/Musical Note Interface worksheet is in two parts to allow a student pair to divide tasks and work simultaneously while transcribing the points from a single graph. Point this out to the students and let them decide who will be responsible for each part of the graph.

STEP 5   Converting polar coordinates to musical notes

Once students have entered the polar coordinates on the coordinate/note interface chart, it is a straightforward process for them to use the number/note reference key to copy the notes that correspond to the polar distances onto the staff. Students do not need to understand the musical notation system to do this step. Some will be curious, and others will know how the notation works. Have the designated musical expert give a brief orientation, or give students the explanation in the following discussion point. In any event, students simply need to copy the notes they see from the key to the staff, paying attention to what line or space the note is on. They should do this in pencil to allow for modification in Step 6.

Musical notes are represented symbolically as dots on a set of five lines called a *staff*. They can be centered on a line or a space. Each line and each space of the staff represents a different note; notes high on the staff sound higher than notes low on the staff. Notes are given letter names, but musicians occasionally refer to them by number names as well. The rhythmic aspect of the notes uses another type of notation; see the activity Measures of Time.

STEP 6   Preparation for performance: Applying rhythm and accompaniment

The notes on the chart are essentially what modern musicians refer to as a *tone row*. These notes have some melodic content by virtue of the fact that they are a specific sequence of pitches, but as written they lack a key element of what creates a melody—rhythm. Students at this point need to decide how the notes will be played rhythmically and what kind of accompaniment will be used. How this step is handled will vary depending on both your musical inclination and that of your students. The activity can be successful even if no rhythm or accompaniment is applied and the notes are played exactly as they sit on the interface chart. Adding rhythm and accompaniment, however, can be fun for students and makes the latter stage of the activity more interesting for the class. Regardless of how far you and your students take this, some performance guidelines need to be adhered to for the melodies to be matched to their images. Go over the following guidelines with your class. Refer to Measures of Time for more information regarding these music basics.

- Each note must be assigned to a beat. The beat of a rhythm is an even pulse in time that all rhythmic events are related to. This assignment must be consistently adhered to by all groups or the melodies will not reflect the images accurately. Consider the fact that successive notes result from successive angles on the graph. If the notes are not consistently matched to beats in time, the resulting melody will imply a different placement of points on the graph and hence a different image.

- Choose a time signature and place bar lines on the staff to establish the rhythmic phrasing. It is okay to let the first few notes be a pickup. (A *pickup* is one or more notes that occur as an incomplete measure before the melody.) Where the first bar lines are placed can alter the rhythmic feel without changing the melodic shape in a way that would distort its representation of the image.

- Choose a designated background rhythm as an accompaniment for the performance. The three rhythmic accompaniments on the CD consist of a drum rhythm played in three styles: rock (track 74), Latin (track 75), and hip-hop/rap (track 76). Use one of these three, or ask a student to bring in a drum to accompany the instruments. The CD rhythms are all in four-four time.

- If notes on the interface chart repeat, you may choose to tie them to create a long note. This will add musical interest and shape to the melody while maintaining the representation of the image.

- Any type of dynamic shape or articulation can be applied as long as it does not distort the time value of the notes outside these guidelines. *Dynamics* in music is the relative volume of notes. *Articulation* refers to how specific notes and rhythms are played: staccato (short and clipped), legato (smooth and flowing), and combinations of these two. Experiment with your own combinations.

- Adding harmonic accompaniment (chords) from a guitar or piano can make the selections more musical and more fun to listen to, but don't let the accompaniment hide the melody. These chords would be a harmonious match for the notes used in the activity: C, Dmin, Emin, F, G7, Amin, and Bdim.

 Students may be inclined to alter the melody rhythmically and embellish it to a point where it is not identifiable. To avoid squelching their creativity, suggest that they prepare two versions: one in which the melody is relatively pure so the class will have a good chance to match it to its image source, and another in which the melody is altered as much as they like. More extensive alterations and compositions are suggested as extension projects.

STEP 7   The classroom concert and image matching

Assemble a core group of willing musicians as a class band or select an individual to play the melodies created.

The graphic images need to be displayed for the class to view as they listen to the melodies. Students can copy their images on the chalkboard, draw them on chart paper, or trace them on overhead transparencies. Making enlarged versions for the chalkboard or chart paper can be problematic if the resulting images are not very accurate and thus no longer correspond to the melody. The advantage of using the chalkboard is that most of the images can be viewed simultaneously and compared. Tracing on overheads allows for much greater accuracy, but flipping through ten or so slides for each melody to try to find a match can be tedious.

When a melody is played and students are trying to match it to the images, coach them to visualize the melody on the polar coordinate graph. Imagine a rotating ray around the pole to be the spoke on a bicycle wheel that moves 10 degrees with each beat of the music. Visualize the length of the ray changing as the pitch changes.

Ask students:
What part of the polar coordinates reflects time? [The angle.]
What part reflects the pitch of the note? [The polar distance.]

If students have difficulty, it can be helpful to talk through some of the examples in Step 8 to stimulate their visualization skills with familiar figures.

STEP 8   Discussion and exercises

The visualization exercises listed here can be done before the performances to enhance students' visualization skills with the interface, or they can be done afterward to assess how students have assimilated the ideas. It can be fun to have students attempt to sing the geometric figures. Consider presenting these exercises in the reverse application, giving students the sound and having them describe the geometric figure.

- What would be the sound of a circle with its center at the pole? [One long note, not changing in pitch.]

- How would the circle sound if its center were moved from the pole? [A smooth rise and fall in pitch over time.]

(continued)

- What would be the sound of a spiral starting at the pole? [A steadily increasing pitch, like playing a scale.]

- What would a square sound like? [Steady rise and fall in pitch.]

- What would a gear shape sound like? [Alternating pitches.]

- What would a star sound like? [Several scale steps up, then back, repeating.]

FOLLOW-UP ACTIVITIES

Textbook assignments

Follow Inside Out with textbook assignments that you would ordinarily use to follow an introduction to the polar coordinate system.

Writing prompts

- Did some images make better melodies than others? Which ones? Why?

- What types of images would translate well to images in this system?

- Did the mathematical system of polar coordinates work well to convey the image? Could music be written in another way that would reflect the image better? If so, how?

- Did you enjoy the activity? What did you learn?

Extensions

- Make music from shapes generated by standard polar equations, such as:

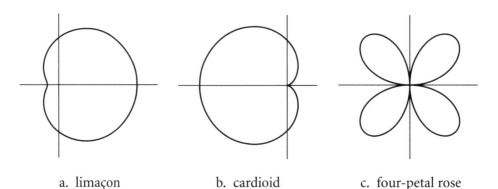

a. limaçon   b. cardioid   c. four-petal rose

- Make images from popular well-known melodies using the system of Inside Out, and let others attempt to determine the melody from the design.

- Use a Cartesian coordinate system to convert images to melodies. Included with this activity is a graph/note interface that can be used to convert images to melodic lines using a rectangular coordinate system. You can present this to students as a creative problem-solving project. Students will need to determine what scale to use for the axes, where to place the axes, which intersection points to use to best convey the image, and what note duration to assign to the horizontal axis units. An example of a mountain range converted to notes is supplied for reference. This example is fairly complex. Simpler versions can be done using only quarter notes. The curve created on the graph can then be represented mathematically as equations of lines or other curves over designated domain intervals. Students can compare the musical lines and mathematical representations in a class presentation.

Project

Using a melody created from an image, students can create variations using techniques from Functional Composer and write a musical composition. When students perform it for the class, ask them to explain why they chose the transformations they used and why those transformations sound good in terms of the mathematical relationships.

TEACHER NOTES

ANSWERS

Polar Coordinate/Musical Note Interface: 0°–180°
Fish Example

Coordinate/Note Interface Chart

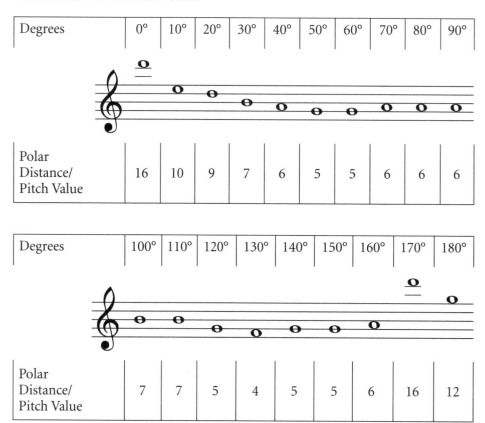

Degrees	0°	10°	20°	30°	40°	50°	60°	70°	80°	90°
Polar Distance/ Pitch Value	16	10	9	7	6	5	5	6	6	6

Degrees	100°	110°	120°	130°	140°	150°	160°	170°	180°
Polar Distance/ Pitch Value	7	7	5	4	5	5	6	16	12

POLAR COORDINATE GRAPH

Fish Example

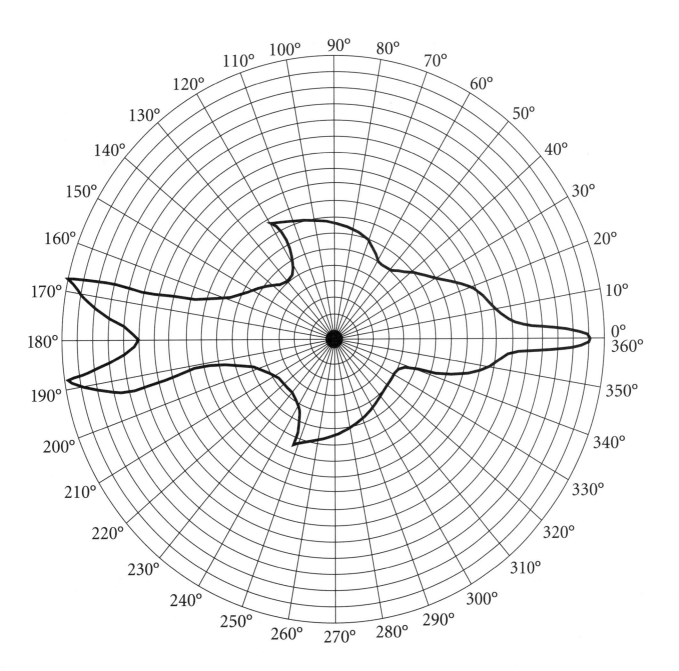

RECTANGULAR COORDINATE/MUSICAL NOTE INTERFACE

Mountain Example

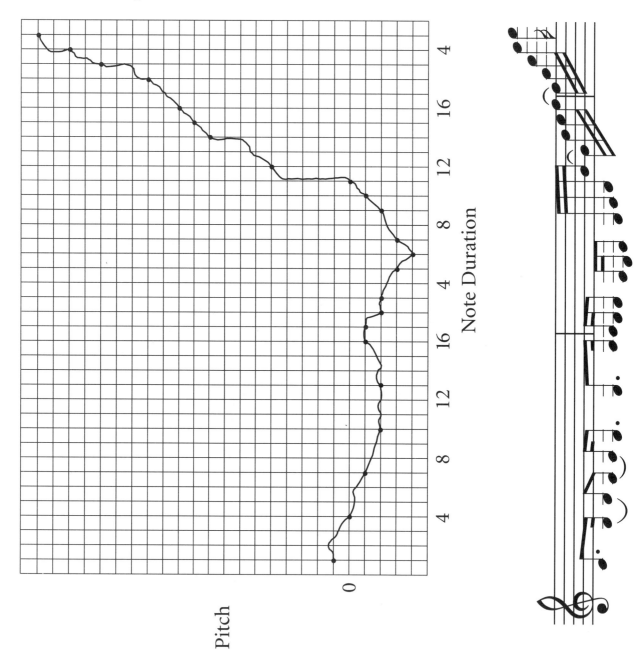

FROM SIGHT TO SOUND

Artists of all types—musicians, poets, painters, sculptors—sometimes get their inspiration from abstract sources that don't seem to have anything to do with their art. Composers write music to communicate things that don't have sound, such as emotions, colors, or natural landscapes. For example, the record "Dark Side of the Moon" by the rock band Pink Floyd uses music to create images of planets rotating in space. Painters often try to express visually things that can't be seen, such as smells, sounds, or emotions. What are these artists thinking or doing to create these effects?

Let's look at painting, for example. If a painting is successful as a piece of art, it can make us feel a wide range of emotions—excitement, peace, anger, frustration, love, victory, despair, you name it. It can even make us feel as if we are in a faraway place and bring to mind all the sights and sounds of that place. In many cases the way this is accomplished is abstract and difficult to analyze. Painters learn their craft, experience the subject, and then express themselves on the canvas—and somehow, the pictures have the same emotions or feelings as the subject. What do you think would happen if a musician did the same thing? Suppose a musician wrote a piece of music and a painter painted a picture, each depicting a fish. What do you think the painting and the music would have in common? The painting may be straightforward—it would look something like the fish. But what about the music? It's natural to try to find a system that makes us think of a fish when we hear the music. We want the sound of the music to fit the fish somehow. Is there some process that can be used,

other than just feeling and interpretation, to make this happen?

If we look deeper, we can see that the fish has a basic structure that can be measured and transferred to both the music and the painting. In the activity Inside Out, you will use a mathematical interface to connect the world of sight to the world of sound. An *interface* is something that acts as a common link between two different systems, sort of like a translator who helps two people who speak different languages understand each other. In a way, our example involves saying *fish* in two different languages: one with music (sound) and one with a drawing (sight). An interface can also be thought of as a secret code. The shape of the fish can be measured and converted to a code. The code can then be converted to music so that the music has many of the qualities of the fish.

You might be wondering whether artists really use mathematics as an interface in this way. Quite simply, some do, some don't, and some do without being aware of it! Many would argue that using mathematics to compose music this way is artificial and lacking in soul. Other musicians have devoted their lives to developing mathematical methods for writing music. For the most part, artists of all types swing between thinking mathematically or technically and working from pure inspiration when they create. In Inside Out, you will explore a mathematical approach. You will experience a way that mathematics can reveal the music contained within the forms of physical objects.

INFORMATION CENTER

Choose a graphic image.

Choose an image that has a distinct contour and can fit comfortably onto the Polar Coordinate Graph.

Trace your image onto the Polar Coordinate Graph.

Center your image so that the center of the graph (the *pole*) is near the center of your image and trace it onto the Polar Coordinate Graph.

Determine the polar coordinates for the image.

Now you need to determine the polar coordinates of each point at which your image intersects the angles of the Polar Coordinate Graph.

- Review of polar coordinates
  To represent any point in polar coordinates, first find the *polar distance:* how far it is from the *pole,* the vertex of all the angles. Then identify the angle the point is on. (Angles are measured in a counterclockwise direction from the horizontal line that corresponds to the positive *x*-axis in the Cartesian coordinate system.) The polar coordinates for the point are given by an ordered pair, with the distance from the pole written first and then the angle.

- Establishing a scale for the polar distances
  Determine what distance from the center each circle on the polar graph will represent. First look at the number/note reference key on your Polar Coordinate/Musical Note Interface worksheets. Use this key to turn the polar distances into notes. What is the highest number you see on the key? Now examine your image on the Polar Coordinate Graph and locate the point farthest from the center. How many circles from the center is this point? Based on this information, decide what polar distance each circle should represent.

- Dividing duties within your group
  Let one person in your group find the polar distances for every 10 degrees from 0 to 180 degrees and another find the distances from 190 to 360 degrees. Note the distance of your image from the pole along the 0-degree line; record that distance, rounded to the nearest whole number, under "0°" on the interface worksheet. This number is the first note of the melody. The number corresponding to the second note is determined by the point at which the image

Information Center (continued)

outline intersects the 10-degree line. Transfer the distances for all intersection points to the space below the appropriate angles on the coordinate/note interface chart.

Convert the polar coordinates to musical notes.

Musical notes can be thought of like numbers on a number line. For this activity each note has been assigned a number. The number/note reference key shows the musical notation for notes corresponding to polar distances. Draw the corresponding note on the staff directly above each polar distance.

Give the notes rhythm and accompaniment.

The notes for your image could be played right off the chart as written, but will be more interesting if you give them a rhythm and an accompaniment. Your teacher will explain guidelines for this.

POLAR COORDINATE/MUSICAL NOTE INTERFACE: 0°–180°

Number/Note Reference Key

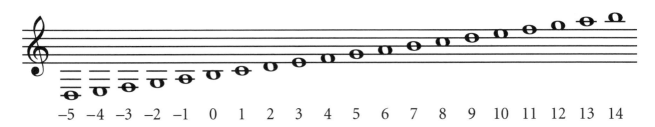

Coordinate/Note Interface Chart

Degrees	0°	10°	20°	30°	40°	50°	60°	70°	80°	90°
Polar Distance/ Pitch Value										

Degrees	100°	110°	120°	130°	140°	150°	160°	170°	180°
Polar Distance/ Pitch Value									

POLAR COORDINATE/MUSICAL NOTE INTERFACE: 190°–360°

Number/Note Reference Key

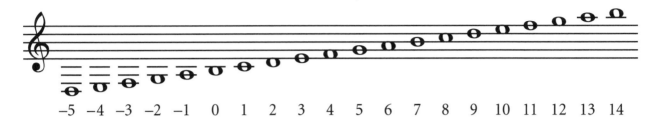

| | –5 | –4 | –3 | –2 | –1 | 0 | 1 | 2 | 3 | 4 | 5 | 6 | 7 | 8 | 9 | 10 | 11 | 12 | 13 | 14 |

Coordinate/Note Interface Chart

Degrees	190°	200°	210°	220°	230°	240°	250°	260°	270°
Polar Distance/ Pitch Value									

Degrees	280°	290°	300°	310°	320°	330°	340°	350°	360°
Polar Distance/ Pitch Value									

POLAR COORDINATE GRAPH

Trace your image onto the graph below. Try to center your image so that the center of the graph is at the center of your image.

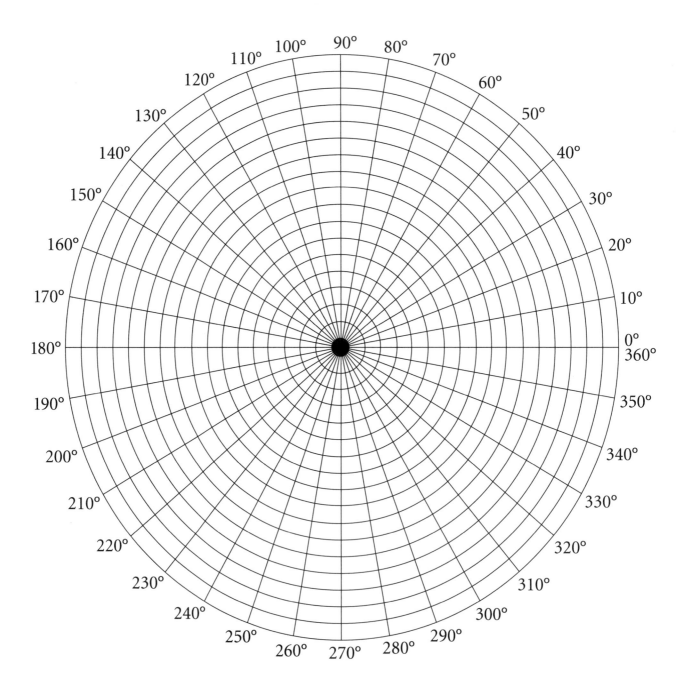

RECTANGULAR COORDINATE/MUSICAL NOTE INTERFACE

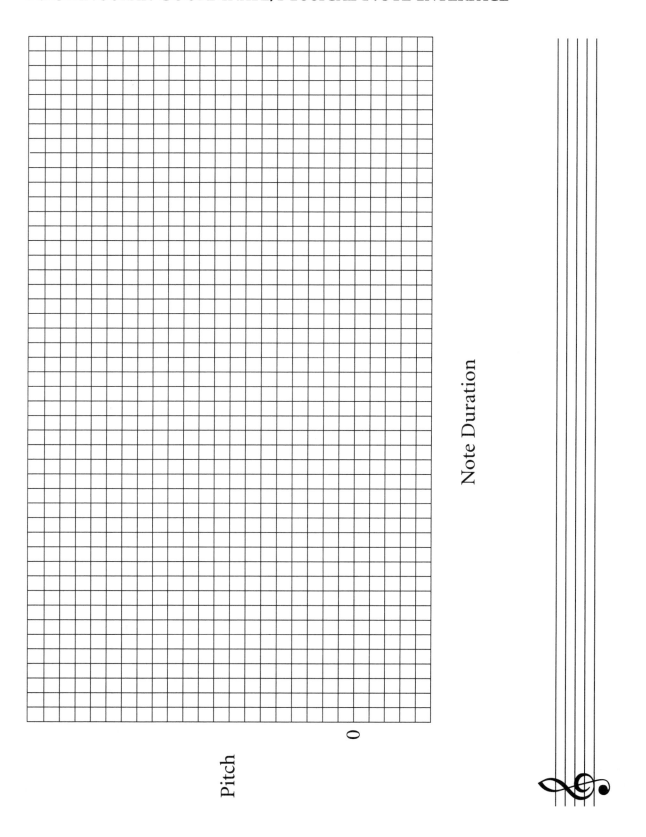

10

Scaling the Scale, Part I

Natural Vibrations and Pythagorean Tuning

Scaling the Scale, Part I leads students through an exploration of how the Pythagorean tuning for the Western major scale is derived from the natural vibrating harmonic frequencies in nature. Students begin by listening to various modes of standing-wave vibrations on a guitar string and seeing a diagram of their vibrating patterns. In a listening exercise, students hear the vibration frequency of each mode in relation to the fundamental and determine the corresponding note in the major scale. The musical distances between the various harmonic tones are defined and named as musical intervals. Students then establish the frequency ratios for the intervals between four of the harmonic tones. Finally, they use the frequency ratios for a fifth and an octave to calculate the frequency ratios for all seven tones of the major scale. This is the method used by Pythagoras to create the tones of the major scale in the sixth century B.C.

Scaling the Scale, Part II will lead students to discover an idiosyncrasy in the Pythagorean tuning. They solve the problem as Marin Mersenne did more than three hundred fifty years ago, using a geometric sequence to create the even-tempered scale—the scale used in modern fixed-pitch instruments.

Mathematics topics

Ratio, operations on fractions (multiplication, division, powers), problem solving, the work of Pythagoras. *Prerequisites:* Fraction multiplication and division, ratios.

Music topics

Scales, intervals, pitch recognition, frequency of pitch, harmonic series, timbre of musical instruments, piano keyboard. *No prerequisites.*

Use with the primary curriculum

- To provide a historical perspective
  When the work of Pythagoras is being studied, use Scaling the Scale to provide an example of the diversity and scope of this mathematician's work.

- To show a connection to physics
  Use Scaling the Scale to provide a graceful and authentic integration point between the physics of sound and mathematics.

- To review fractions
  Use Scaling the Scale to review operations on fractions.

Objectives

- To apply and reinforce fraction concepts
  Nonroutine problems can reveal gaps in basic skills and weaknesses in understanding. This context provides a rich, memorable experience to enhance retention and understanding of fractions.

- To use mathematics as a problem-solving tool in a nonroutine context
  Problem-solving creativity is enhanced with practice in nonroutine problem contexts.

- To see the mathematical and physical aspects of music
  The building blocks of music are derived from natural laws, and these natural laws adhere to an elegant and simple mathematical structure.

- To realize that Pythagoras concerned himself with more than just right triangles

Student handouts

- Natural Vibrations (reading; one per student)

- What to Do (resource page; one per group)

- Standing-Wave Vibrations (worksheet; one per student)
- Notes As Vibrations (worksheet; one per pair)

Materials

- CD tracks 77–88
- Ten-foot length of rope or heavy string (optional)

Instructional time

50–90 minutes

Instructional format

Scaling the Scale engages students in a wide range of topics in a relatively condensed activity. For this design to work, the music aspects have been simplified to the essentials. Take some time to assess the musical background of your class and note the discussion points in the teacher script that indicate where these simplifications take place. You may want to adjust the activity to include discussion of the music topics. Some of the beginning exercises may seem too simplistic for some of your students. Adjust your pace accordingly.

You will use the CD tracks as you lead the first half of the activity and help students understand some music concepts, terms, definitions, and basic mathematical relationships. (A summary of these concepts is in Step 1 from Scaling the Scale, Part II on pages 161–163.) Halfway through the activity, the format shifts to student groups working independently. Take the liberty of supplying instruction during this part of the activity, depending on the needs of your particular class.

Student preparation

Assign or read Natural Vibrations and hold a brief discussion emphasizing the idea that the laws of nature determine the special ways that things vibrate.

ACTIVITY SCRIPT

STEP 1　The natural modes of vibration of a guitar string

The activity begins with an examination of the natural vibrations of a stretched guitar string. Students will examine a variety of different natural modes of vibration for the string. It is important to note that all the modes they will study occur on a string of constant length and tension. The only changes in sound result from the way the string is vibrating.

The first mode is the simplest: The string is plucked with no fingering or intervention of any kind, and a tone sounds. CD track 77 is an example of this. This note is called the *fundamental* or *first harmonic,* because it is created from the primary, most basic way a string can vibrate. The open guitar string vibrates back and forth for its full length, as shown on the diagram on the Standing-Wave Vibrations worksheet. The guitar string heard on the recording is vibrating at 130 Hz. It is called a *C note.* This activity will determine the frequencies of five different modes of vibration. Students should record on their worksheets the frequency of each mode both in terms of its actual value as heard on the CD and as multiples of the starting frequency, *f.*

Students see the first frequency values filled in on the student worksheet.

Steps 1 to 3 may be very basic for some students with music backgrounds. You can adjust accordingly how much time you spend developing and discussing the harmonics.

STEP 2   Another mode of vibration

Explain that if a guitarist places a finger directly next to the string when it is plucked and instantly removes it, the obstruction of the finger forces the string to vibrate differently, pivoting around the center point as shown in the diagram on the Standing-Wave Vibrations worksheet. This creates the second harmonic. The pivot point, called a *node,* divides the string exactly in half. The node does not move, while the sections of string on either side vibrate back and forth in opposite directions.

Different modes of vibration can be effectively demonstrated by using a rope or a string. With one person holding each end of the rope, the first and second vibration patterns can be simulated. Simulating the third is almost impossible, but it can be fun for students to try.

Play CD track 78 and direct students' attention to the vibration diagram on the worksheet.

Ask students:
Can you describe the difference in sound between this note and the first harmonic? [One sounds higher than the other, maybe twice as high.]
What do you think is the relationship between the frequency of the first harmonic and that of the second harmonic? Refer to the hints on the resource page. [The second harmonic is vibrating twice as fast.]

Have students record these values in the appropriate space on their worksheets.

STEP 3   Frequencies of the remaining harmonics

Using the method you used with the first two harmonics, establish the frequencies for the third, fourth, and fifth harmonics. Play the CD track for each one (track 79 for the third harmonic, track 80 for the fourth harmonic, and track 81 for the fifth harmonic). Ask students to determine the frequencies and fill in their worksheets.

Strings vibrate in modes well beyond the fifth harmonic. All of the harmonics are referred to in music as the *harmonic series*. In mathematics, the term *harmonic series* refers to the nonconverging infinite sum

$$1 + \frac{1}{2} + \frac{1}{3} + \frac{1}{4} + \frac{1}{5} + \frac{1}{6} + \frac{1}{7} + \frac{1}{8} + \frac{1}{9} + \dots$$

The terms of the mathematical harmonic series are the same as the fractions representing the vibrating string length between nodes for each successive harmonic. The second harmonic is $\frac{1}{2}$ the vibrating length of the fundamental, the third is $\frac{1}{3}$ the vibrating length, and so forth. While musical string lengths in the musical harmonic series are not being added to create a sum, the mathematical counterpart does suggest the question of whether there are an infinite number of harmonics. [Theoretically, yes; in practice, no.]

STEP 4   Reflection on the harmonic series

Ask students:

Do any of these notes sound like notes from music that you have heard?

Wind instruments as well as string instruments are subject to the same laws of nature as guitar strings. They create the same harmonic notes naturally. A bugle has no valves, so the length of its tubing is fixed in the same way that the guitar string's length remained fixed as it was placed in different modes of vibration. For a fixed length, the only notes that can sound are the notes of the harmonic series.

Recall bugle melodies that students may know, such as "Taps" or "Reveille." Ask a student volunteer to sing or hum them, or hum them yourself.

Some classic bugle tunes are played using harmonics on the guitar on track 82. Play the track.

The notes above the first harmonic are also referred to as *overtones*. Though not individually identifiable, many of these overtones above the fundamental frequency (the first harmonic) sound when an instrument plays a single note. In essence, the vibrating medium of an instrument actually vibrates in several modes simultaneously. The relative loudness of the overtones is what gives any note its characteristic sound—the way a trumpet, for example, can be distinguished from a guitar or a piano. The sound of an instrument is called its *timbre.* The relationship of the intensity of various overtones affects the timbre of a musical instrument. The timbre of a guitar string (the relative loudness of its various harmonics) can be changed by plucking the string at different places along its length, generating different combinations of vibration modes.

STEP 5   Harmonic tones and the major scale

This step of the activity will use six tracks of the CD. In a listening exercise, students will locate the notes created by the harmonic series in the major scale.

This exercise could be skipped to save time. You can give the positions of the notes as indicated on the teacher answer key directly to students.

To familiarize students with the major scale, explain that most music uses groups of notes called *scales.* The notes are given letter names and number names according to their position in the scale. Turn your attention to the piano diagram on the Standing-Wave Vibrations worksheet. Point out that the pattern on the keyboard repeats after seven notes and that the numbering reflects this. The eighth tone of the first pattern (octave) is the first tone of the next pattern. Play track 83 and ask students to follow along on the keyboard as the scale is played and the note numbers are announced on the CD. This activity should familiarize nonmusicians with the idea of scales.

This activity will consider only the white keys, the *diatonic tones* of the C-major scale. The black keys are notes that fall between the white keys. They create half steps and combine with the white keys to make up the *chromatic scale.* Tell students to ignore the black keys for now. They will consider them in Scaling the Scale, Part II.

Once students are familiar with the major scale, their next task is to learn to locate the harmonic tones.

Tracks 84–88 on the CD focus on each harmonic again, but in relation to the major scale. Each track sounds a different harmonic together with the C-major scale. Have students follow along on the piano keyboard as the scale sounds and place an **X** below the key that has the same pitch as the harmonic that is sounding. They should place the frequency for the harmonic below the **X** (in terms of the general value, f) at this time as well. Work your way through the tracks, pausing after each one to briefly discuss what was heard. The first harmonic is on track 84, the second is on track 85, and so on to the fifth harmonic on track 88.

Some students with musical experience will be able to locate the harmonics on the scale without the CD tracks. Others may struggle with what they hear on these tracks. Don't let them get bogged down. Coach students and, as a class, come to a conclusion as to the correct placements.

STEP 6 Musical intervals between harmonic tones

At this point you will teach the concept of a musical interval.

The activity may be successful with minimal discussion of intervals. You can make this decision based on your own comfort level, the amount of time you have, and the background of your students. The minimal working understanding of intervals for this activity is presented on the student resource page.

Explain to students that a musical interval is related to the distance between two notes or piano keys. (Any reference to a piano key is to white keys only. Remind students that they are ignoring the black keys for now.) The naming of the intervals is very logical. The name for the interval between tone number 1 and tone number 8 is an *octave* (from the Latin *octava*, meaning "eighth part"). The name for the interval between tone number 1 and tone number 5 is a *fifth*. The interval between tone number 1 and tone number 4 is a *fourth*, and so on. This pattern is true for all white keys on the piano. If you choose any white key and call it number 1, the interval created by going one white key up or down is a *second*; two white keys up or down, a *third*; and so on. The resource page simplifies this idea in a way that is functional for the activity.

This system may cause some confusion when you consider an interval as a distance. In a musical interval, a second is a distance of 1, a third is a distance of 2, and so on. This situation can be generalized as

$$\text{musical interval name} = \text{mathematical distance} + 1$$

As with many musical topics in this book, this treatment of musical intervals has been simplified to streamline the activity for nonmusicians. The interval names second, third, fourth, and so on are general terms. Each can be designated more specifically as minor/major second, minor/major third, perfect or augmented fourth, and so on, each term representing a different musical distance. If you examine the piano keyboard, it is clear that there is no black key between notes 3 and 4 and notes 7 and 8. This suggests that the distance between notes 3 and 4 and the distance between notes 7 and 8 are smaller than the distance between the other white keys. This is true. However, using the general names is a simplified and accurate system of naming the intervals in relation to movement between any white keys. If your musical students bring this up, acknowledge it and explain that, for the purposes of the activity, this point is not essential. In fact, it may confuse the matter and be distracting to your goal. You should refer to the intervals by their general names. Technically, however, the intervals between the harmonic tones considered in this activity are the octave, perfect fifth, perfect fourth, and major third.

On their worksheets, have students label the musical intervals between the harmonic tones with the appropriate names. The appropriate names are shown on the teacher answer key.

STEP 7   Frequency ratios for musical intervals

This step develops ratios between the frequencies of the notes of an octave, fifth, fourth, and third (the first four overtones).

Help students establish a frequency ratio for each interval created between adjacent harmonic tones. Explain to them that the object is to find a factor that will allow them to calculate the frequency of a note a given interval above the given note. For the interval of a fifth, we can let the factor be x, so $2f$ times x equals $3f$. Solving for x gives a factor of $\frac{3}{2}$. To create the interval of a fifth, multiply a note's frequency by $\frac{3}{2}$ to get the frequency of the higher note.

Students will indicate these ratios on the chart on the Standing-Wave Vibrations worksheet. As you teach this step, keep in mind that a musical interval name is the mathematical distance plus 1.

When you establish the ratios for the musical intervals from the harmonic tones, it is important to keep in mind that the ratios apply for those intervals (distances) wherever they may occur on the piano keyboard.

STEP 8   Adding intervals and creating the major scale

Give student groups the second worksheet, Notes As Vibrations. As students work, you can move from group to group, coaching and clarifying when necessary. Alternatively, you can complete the worksheet as a class.

To get groups started, clarify that laying intervals end on end should be done in such a way that the top note of the first interval is the bottom note of the second interval. That is, the intervals share a common note.

Students may need some assistance in finding a process by which to calculate the ratios of the scale tones. The instructions are left purposely vague to give students a greater opportunity to explore and discover. Direct them to the hints on their worksheet and provide direct instruction when absolutely necessary.

If students have begun working on the development of the scale (the frequency ratio to fundamental), they can finish the chart for homework. It is important for students to be sure their answers to the first problems are correct, since each problem is dependent on the preceding answer.

STEP 9   Closure: Ratios for Pythagorean tuning of the major scale

Bring the class together and check the ratios for each tone of the major scale. Explain that the Pythagorean tuning is only one of many that have been used through history. Early scale tunings, such as the Pythagorean tuning, presented problems for musicians as music became more sophisticated and different keys were used. In Pythagorean tuning, musical intervals of the same name (such as seconds or whole steps) that occur at different places in the scale actually have different ratios. This idiosyncrasy was finally solved with the tuning method currently used as a standard for Western music, called *even temperament*. Even temperament is presented in the activity Scaling the Scale, Part II.

 Notice that the frequency ratio for the third created by the harmonic tones is $\frac{5}{4}$, while the frequency ratio for the third calculated in the activity is $\frac{81}{64}$. This is one example of the idiosyncrasies of Pythagorean tuning. The third from the harmonics is a pure tuning, while the third from the activity is the Pythagorean tuning. The ratio for the pure third is actually smaller than for the Pythagorean third. Many musicians contend that the pure third (natural) is more pleasing than the Pythagorean or tempered third (calculated). Another type of tuning, called *just intonation,* substitutes the ratio for the pure third into the Pythagorean tuning. The debate over pure tunings versus mathematically altered tunings continues to this day among some musicians.

FOLLOW-UP ACTIVITIES

Scaling the Scale, Part II

Scaling the Scale, Part II reveals an idiosyncrasy of stacking intervals and multiplying frequency ratios. Chromatic notes are introduced, and students solve the centuries-old problem of tuning the scale by creating the system currently used in Western music: even temperament.

TEACHER NOTES

Writing prompts

- What did you learn in today's activity?

- What part of music is a human invention, and what part is determined by nature?

- Is it surprising to you that Pythagoras was so involved in solving musical problems? Why or why not?

Textbook assignments

- Textbook problems that apply operations with fractions, arithmetically or algebraically

- Textbook problems that use the Pythagorean theorem

Extension

Challenge students to find ratios for the scale tones by applying the same interval stacking process using the ratios of the fourth and the third of the harmonic tones in addition to those of the fifth and the octave. The pure-scale tuning can be developed using these ratios.

ANSWERS

Standing-Wave Vibrations

	Frequency	
	Example	General
First harmonic	130	f
Second harmonic	260	$2f$
Third harmonic	390	$3f$
Fourth harmonic	520	$4f$
Fifth harmonic	650	$5f$

C	D	E	F	G	A	B	C	D	E	F	G	A	B	C	D	E	F	G

Harmonic frequency	f						2f			3f		4f	5f
Musical interval		Octave					Fifth			Fourth		Third	
Ratio for interval		$\frac{2}{1}$					$\frac{3}{2}$			$\frac{4}{3}$		$\frac{5}{4}$	

Notes As Vibrations

1. A fifth and a fourth
2. $\frac{3}{2}$ and $\frac{4}{3}$
3. $\frac{3}{2}$ times $\frac{4}{3}$ equals $\frac{2}{1}$
4.

Musical interval	Frequency ratio to fundamental
The fundamental	1
The second *up 2 fifths, down 1 octave* $\left(\frac{3}{2}\right)\left(\frac{3}{2}\right)\left(\frac{1}{2}\right) = \frac{9}{8}$	$\frac{9}{8}$
The third *up 4 fifths, down 2 octaves* $\left(\frac{3}{2}\right)\left(\frac{3}{2}\right)\left(\frac{3}{2}\right)\left(\frac{3}{2}\right)\left(\frac{1}{2}\right)\left(\frac{1}{2}\right) = \left(\frac{3}{2}\right)^4\left(\frac{1}{2}\right)^2 = \left(\frac{81}{64}\right)$	$\frac{81}{64}$
The fourth, frequency of the natural harmonic	$\frac{4}{3}$
The fifth, frequency of the natural harmonic	$\frac{3}{2}$
The sixth *up 3 fifths, down 1 octave* $\left(\frac{3}{2}\right)\left(\frac{3}{2}\right)\left(\frac{3}{2}\right)\left(\frac{1}{2}\right) = \left(\frac{3}{2}\right)^3\left(\frac{1}{2}\right) = \left(\frac{27}{16}\right)$	$\frac{27}{16}$
The seventh *up 5 fifths, down 2 octaves* $\left(\frac{3}{2}\right)\left(\frac{3}{2}\right)\left(\frac{3}{2}\right)\left(\frac{3}{2}\right)\left(\frac{3}{2}\right)\left(\frac{1}{2}\right)\left(\frac{1}{2}\right) = \left(\frac{3}{2}\right)^5\left(\frac{1}{2}\right)^2 = \left(\frac{243}{128}\right)$	$\frac{243}{128}$

NATURAL VIBRATIONS

Have you ever wondered where the organization of musical notes comes from? Every day we hear many sounds: a bee buzzing, car brakes squeaking, people humming tunes, wind whistling through the trees. These sounds can make us feel happy or sad, excited or calm. They affect us in many ways. The sounds we call music are unique in that they are intended to change the way we feel. What are sounds, and what are notes?

In fact, music is a series of notes, and notes, like all sounds, are just vibrations. The rate or *frequency* of vibration determines what note we hear. A fast vibration (high frequency) is a "high" note, and a slow vibration (low frequency) is a "low" note. Musicians organize these different frequencies of vibration in ways that our ears find pleasing—in ways that make music.

For musicians to make music that sounds good to our ears, the frequencies must have some kind of organization. You can't just string together a bunch of vibrations and think it will be pleasing. So where does the organization of vibrations come from? Is it natural, or is it entirely a human invention?

Have you ever heard the rope on a flagpole bang around in the wind? It seems random, but after a gust of wind, the way the rope bangs against the pole always follows the same pattern. In this way, the rope on the flagpole is like a string on a guitar. If you pick up a guitar and pluck a string, the string will definitely vibrate at a particular frequency. The moving string makes the air vibrate with the same frequency. The vibrating air then strikes your ear, and you hear one pure note—the note that corresponds to the frequency of the guitar string's vibration.

Guitar players make different notes by shortening and lengthening the string with their fingers, pressing the string against the fretboard. Short strings vibrate faster than longer strings, so short strings make higher notes than longer strings. Another way guitarists can change the frequency of vibration (the note) of a guitar string is to change how tightly the string is stretched. Strings with more tension have a higher pitch than strings with less tension. A third way to change the frequency is to change the material the string is made of.

When we look deeper, we see that any string has a variety of different but related ways it can vibrate without anyone changing its length, its tension, or what it's made of. These ways of vibrating produce different notes, called *harmonics*. Western classical musicians organized these harmonics into *scales* and used them to make music.

The harmonic vibrations of a string follow a natural law of physics that humans did not create and cannot change. So, as it turns out, the basis for the music we hear today was created by nature.

The ancient Greek mathematician Pythagoras recognized this. He studied the mathematical relationships between vibrating strings and pitch and developed a way to determine the frequencies that have formed a basis for Western music. In the activity Scaling the Scale, you will follow the same steps that Pythagoras followed—going from observing nature's vibrations to creating a Western musical scale.

WHAT TO DO

Listen and guess.

Your teacher will play a recording of the first and second harmonics. Listen to the difference in pitch and look at their vibration patterns on the Standing-Wave Vibrations worksheet. Discuss what you hear and see, and make a logical guess as to what the frequency of the second harmonic is in relation to the first.

Hints

- Higher-sounding notes vibrate with higher frequencies.
- The node of the second harmonic is exactly in the middle of the string.

Calculate harmonic frequencies.

Using the pattern you observed, fill in on your worksheet the frequencies for the rest of the harmonics.

Find the harmonic frequencies on the piano.

Listen to the examples played by your teacher and place an X directly below the piano key that sounds like the same note as the harmonic you hear. Repeat this process until you have located all of the harmonics on the piano keyboard. Below each X on the diagram, write the general frequency value for each harmonic in terms of the fundamental, f.

Understand musical intervals.

Musicians call the distance between two notes an *interval* and have special names for them. Use the chart below to find the names of the intervals between each pair of harmonics. Write these names in the table on your worksheet.

Interval name	Keys on piano
Second	spans 2 white keys
Third	spans 3 white keys
Fourth	spans 4 white keys
Fifth	spans 5 white keys

Find the ratio for the interval.

Calculate the frequency ratio between the top note and the bottom note of each interval and place it on the table on your worksheet.

Create the major scale.

Follow the steps Pythagoras followed as you complete the Notes As Vibrations worksheet and find the frequency ratio of each note to the fundamental.

STANDING-WAVE VIBRATIONS

	Frequency		Vibrating string pattern
	Example	General	
First harmonic	130	f	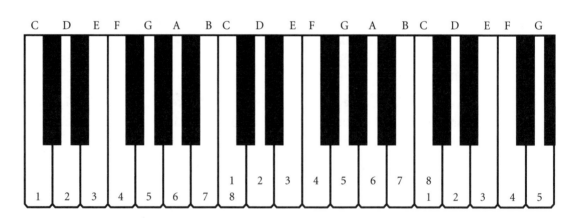
Second harmonic			
Third harmonic			
Fourth harmonic			
Fifth harmonic			

Harmonic frequency	
Musical interval	
Ratio for interval	

NOTES AS VIBRATIONS

You have just learned how the natural vibrating frequencies occur in the musical scale used by modern Western music. The exploration you are about to make develops the major scale. It was followed by Pythagoras about 2,500 years ago.

1. Refer to the diagram of the piano keyboard. If you were to lay intervals end to end (with the last note of one overlapping the first note of the next), what two intervals of harmonic tones would fit exactly within one octave?

2. Now consider the frequency ratios for these two intervals and write them here.

3. An octave has a ratio of $\frac{2}{1}$. What mathematical operation (addition, subtraction, multiplication, division) between the two intervals you found in the first step and whose interval ratios you listed in the second step gives $\frac{2}{1}$? Experiment. Show your work.

4. Your task is to find the frequency ratios of the other four notes of the major scale relative to the first tone. Use the operation you just discovered. Like Pythagoras, you may use only ratios of the fifth and the octave. You do not need to calculate the intervals for the fourth and fifth. Pythagoras used the natural harmonics for these intervals. Show your work in the spaces provided in the table on the following page.

Hints

- Start with tone number 1. Experiment to see if you can move up the piano by some number of fifths or octaves and then down by another number of fifths or octaves to land on one of the tones you are trying to find. If you find a combination that works, you can use operations with the frequency ratios to find the frequency ratio for the final note.

- To obtain the frequency of a note a fifth higher than a given note, multiply the ratio for that note by $\frac{3}{2}$.

Notes As Vibrations (continued)

- To obtain the frequency of a note a fifth lower than a given note, divide the ratio for that note by $\frac{3}{2}$.

- The same process works for the octave: Multiply by 2 to find the frequency of a note an octave higher, and divide by 2 to find the frequency of a note an octave lower.

Musical interval	Frequency ratio to fundamental
The fundamental	1
The second	
The third	
The fourth	$\frac{4}{3}$
The fifth	$\frac{3}{2}$
The sixth	
The seventh	

11

Scaling the Scale, Part II

A Solution to the Limitations of Pythagorean Tuning

In this activity, students confront and solve a conflict between the Pythagorean frequency ratios of the major scale and the way the piano keyboard is organized. The opening introduces students to the black keys of the piano (the chromatic tones) and establishes a new way to define intervals of the fifth and the octave using chromatic tones (also referred to as *half steps*). Students stack intervals of the fifth and the octave over the length of the piano keyboard until some number of fifths lands on the same note as some number of octaves. The ratio for the resulting interval is calculated with fifths and then with octaves using methods from Scaling the Scale, Part I. The two methods, which should yield the same ratio, in fact yield slightly different numbers. Students solve the problem by creating a system in which all of the frequencies of the chromatic scale

are placed in a geometric sequence. This produces the *even-tempered scale*, the current standard for modern instruments. It was first developed by Marin Mersenne more than three hundred fifty years ago to solve the temperament problem and was made famous with Bach's "Well-Tempered Clavier."

This activity is the most theoretical of any in the book; it uses little audio accompaniment. Students engage not directly with music but with the mathematical organization of the scales that are used to create music. Thus the activity tends to appeal more to the academic student and is less effective in reaching visual and audio learners.

The topic of musical temperaments can be very complex and has been carefully simplified in this activity to make the essence of the historical dilemma and the solution of scale tunings accessible to students.

Mathematics topics

Geometric sequence (*n*th term, common ratio, formula), problem solving, patterns, multiples, fraction multiplication, ratios. *Prerequisites:* Knowledge of a geometric sequence and application experience using it to model real-world problems, ability to multiply fractions.

Music topics

Scale temperaments (frequencies of scale tones), chromatic tones, intervals, piano keyboard layout, combining intervals. *Prerequisites:* Scaling the Scale, Part I or knowledge of interval names and definitions, piano keyboard layout, frequency ratios of harmonic series tones.

Use with the primary curriculum

- To assess student understanding of geometric sequences
  After students have studied sequences, use Scaling the Scale, Part II to assess students' ability to apply the concepts in a nonroutine problem.

- To introduce geometric sequences
  Use this activity as a compelling way to introduce the study of a geometric sequence. At the beginning of their study, have students do the Problem with Pythagorean Tuning worksheet to establish a problem. Leave the activity, develop the knowledge and skill necessary to solve the problem, and then return to the activity.

- Between units as a special-interest, interdisciplinary application
  Use Scaling the Scale to engage students in a real-world application of mathematics that crosses curricular boundaries.

Objectives

- To enhance retention
  By doing the mathematics that makes modern music possible, students receive an impression that can enhance their retention of the concepts and skills.

- To motivate
  Music is important to students' lives. Solving musical problems with mathematics makes its study more meaningful.

- To increase awareness of mathematics in history
  When mathematics appears in unusual applications, it inspires interest, curiosity, and appreciation.

- To increase confidence with nonroutine problems

Student handouts

- Natural Vibrations (reading for Part I on page 154; optional)

- Natural Harmonics (resource page; one per group)

- Problem with Pythagorean Tuning (worksheet; one per student)

- Even-Tempered Scale (worksheet; one per student)

Materials

- CD track 89

- Overhead transparency of the Natural Harmonics resource page (optional)

Instructional time

35–50 minutes

Instructional format

During most of this activity, students work in pairs or groups. Your role is coach and facilitator. You will bring the class together for discussion in the middle of the activity (after the first worksheet) and again at the end.

If Scaling the Scale, Part I has not been completed, you will need to provide the necessary background information. Most students will understand this information better if they've spent a day working through the earlier activity.

Student preparation

If Part I has been completed, no extra preparation is needed to begin Part II. Part II follows directly from Part I.

If Part I has not been done, students should read and discuss Natural Vibrations the day before this activity.

ACTIVITY SCRIPT

STEP 1   Review background material

If Scaling the Scale, Part I has been completed, review the ideas from Natural Vibrations and refer to the resource page for this activity as you review the ideas of musical intervals, their frequency ratios, and their natural origins (such as vibrating guitar strings). After this, proceed directly to Step 2.

If Part I has not been completed, you will need to invest about 15 minutes in establishing the music concepts that follow on the next page:

T E A C H E R   N O T E S

TEACHER NOTES

Standing waves   The student reading in Part I referred to the ways that things vibrate. The resource page shows the ways a stretched guitar string vibrates, sounding different notes, without changing the length or tension.

Strings vibrate freely in various modes called *standing waves.* Each graphic on the resource page represents a string's vibration in a different standing-wave mode. The tension and length of the string are constant for all modes.

First harmonic, *f*   This mode, called the *fundamental,* occurs when an open string is plucked. The full length of the string swings back and forth freely. The frequency is determined by the tension and length. We will call this frequency *f.*

Second harmonic, 2*f*   This harmonic is created when an obstruction is placed directly at the midpoint of the string when the string is struck. The presence of the obstruction (the guitarist's finger) forces the string to pivot around the center point in its vibration. The guitarist does not hold the string; the finger is placed next to the string and instantly removed after the string is struck. The pivoting vibration occurs freely, with no further intervention by the guitarist. The pitch of this mode is heard as an octave higher than the fundamental, and its frequency is 2*f,* exactly double that of the fundamental.

Third harmonic, 3*f*   To create the third harmonic, the pivot points divide the string into three equal parts. To create this mode, the guitarist's finger must be placed in a position one third of the distance from the end of the string. As nature and our intuition would have it, the frequency is three times that of the fundamental, or 3*f.*

Remaining harmonics   Each remaining harmonic occurs in the same way, with frequencies neatly organized as consecutive-integer multiples of the fundamental.

   In application, all single pitches created by musical instruments contain some combination of these harmonic frequencies within them. The ear hears the pitch as the fundamental, but the presence of the various harmonic frequencies (also known as "overtones") gives the note its characteristic sound: shrill, warm, thin, and so on. When a guitar string vibrates, it actually vibrates in many of these modes at once.

Musical intervals   Refer to the piano keyboard on the resource page. Each key of the piano represents a different note with a different frequency. Musicians assign the notes number names in addition to letter names, likening the row of notes to a number line that begins at 1. The musical distance between any two notes is called an *interval.* An interval is not a mathematical distance between keys; instead, every key is counted. The movement from one white key to the next is called a *second*. There are three keys between a note and its fifth. The equivalence for the names of musical intervals in relation to number-line distance is

$$\text{music interval} = \text{mathematical distance} + 1$$

The intervals created by the harmonic series are indicated on the resource page.

Notice that the harmonic series shown on the resource page includes the sixth harmonic, which creates the musical interval of a minor third from the fifth harmonic. A minor third is a smaller musical distance than a major third by one half step, giving it a different frequency ratio. Half steps and the chromatic scale are discussed in Step 2.

Frequency ratios The last idea needed before beginning the exploration is the frequency ratio for each musical interval. Notice that the interval of an octave is created by doubling the frequency of a given note. What multiplier would you use to create the frequency of the interval of a fifth? If this multiplier is x, then $2f$ times x equals $3f$ for a fifth, yielding $x = \frac{3}{2}$. All the ratios on the resource page can be found this way.

If you or your students have completed Scaling the Scale, Part I, you may notice a discrepancy between the interval ratios created by the harmonics and those created by the Pythagorean tuning method. Particularly, the ratio for the major third created by harmonics (pure tuning) is $\frac{5}{4}$, whereas the Pythagorean ratio is $\frac{81}{64}$. The ear will hear both of these frequencies as a major third, but the pure third is slightly lower than the Pythagorean third. This is one example of a tuning inconsistency that will be solved in this activity.

This is a cursory overview of the musical concepts necessary for the exploration. Details have been selectively omitted to streamline the process and to avoid distraction from the focus of the activity. Feel free to expand the discussion as you see fit and to adapt it to your background and that of your students.

STEP 2 The black keys: Chromatic tones

Before students begin the worksheet, let them hear the musical scale. Scaling the Scale, Part II presents a dimension of the scale that was not considered in Part I: the black keys. These keys are notes in between the notes of the white keys. The black keys create the *chromatic scale*. The musical interval between any two adjacent keys, white or black, is called a *half step*. The CD track has an example of the scale played once without the chromatics and then once with the chromatics to give students a sense of what these notes add musically.

Because the system of intervals that has been established so far uses only the white keys, many questions may arise. Some of your students will recognize the black keys as sharps and flats. A discussion of these notes could become very complex. Many musical details of chromatics are not essential for the activity and could distract from the focus. The necessary aspects of black keys are explored in the first worksheet. If you are so inclined, however, you could elaborate on some of the musical aspects at this point.

STEP 3   The first worksheet: Problem with Pythagorean Tuning

Students work through the first worksheet on their own. Move from group to group, offer help when it is needed, and make sure groups are on the right track. Each question requires that students correctly answer the question preceding it.

It is essential that students understand what they need to do to answer question 3. They must count seven half steps (this is eight notes of the chromatic scale, both black and white keys, when the starting and ending notes are counted) to cover the interval of a fifth, and twelve half steps for the interval of an octave. The ambiguity resulting from the number of chromatics not being consistent with the actual interval name may disconcert students. Mention that the name is related to the number of white keys.

You may also need to provide students with an example of "laying intervals end on end." Start with the C note indicated on the resource page. A fifth (seven half steps) spans tones 1 to 5. Starting with the ending note of the previous interval and counting up another seven half steps (another fifth) spans tones 5 to 2. Stacking another fifth spans tones 2 to 6, and so on. This chain of intervals can continue up the entire keyboard. It is important to point out that at one point the starting and ending notes will not be white notes. Continue to determine fifths by counting seven half steps, black or white. Octave intervals can be stacked in the same way, counting 12 half steps. The cycle of stacking octaves is very straightforward in its configuration on the keyboard. The object is to find where these cycles match. While students may begin this process by mechanically counting on the keyboard, a quick analysis of the situation reveals that the number of chromatic tones where the cycle of stacked fifths matches the cycle of stacked octaves will be the least common multiple of 7 and 12, which is 84. Thus on the piano keyboard 12 fifths fit exactly into 7 octaves. Let students use their own methods. The counters can gain a deeper understanding of the concept of least common multiples if they follow their own paths.

When most of the students are at this point, bring the class together to discuss the problem discovered with the Pythagorean ratios. After they have answered question 7, bring the class together to acknowledge the dilemma.

The significance of the discrepancy they discovered between Pythagorean ratios and the piano keyboard may be too abstract or obscure to carry great meaning for some students. You may find it valuable to take more time to discuss the implications. This discrepancy indicates that the frequencies used to tune a piano are not derived by the Pythagorean ratios. Why not? Use of the Pythagorean ratios creates slightly different ratios for many intervals of the same name on the piano. Consequently, if a piano were tuned using Pythagorean tuning in relation to C, scales built on other notes (known as different keys) would sound different and out of tune. A real understanding of this idea is beyond the scope of this activity, but the general idea can be established. Musicians struggled for years to find ways to establish the frequencies of the scale whereby all keys sound in tune.

STEP 4   The second worksheet: Even-Tempered Scale

Have a student read aloud to the class the paragraphs that summarize the problem and refer to the historical context. Have students work on the second worksheet in their groups. As before, you may need to clarify, coach, and provide hints.

STEP 5   Closure: Discussion of the final worksheet question

After students have completed question 3 on the second worksheet, which may be assigned as homework, discuss how scale tones of the even-tempered tuning differ from those of the Pythagorean tuning for the major scale. Many of the ratios for even temperament are irrational numbers. Except for the octave, none of the frequency ratios created by even temperament are the same as those for Pythagorean tuning or for the natural harmonics.

Scale tunings have been the subject of controversy historically, and the controversy still persists today.

Even temperament is considered a compromise—a mathematical adjustment of nature's pure harmonies to enable fixed-pitch instruments to play in many keys. Some believe that by making the ratios between all pairs of adjacent notes the same, music has been sterilized and the character of individual keys has been destroyed.

Believing that even temperament lacks the essential beauty present in tunings that use the ratios of the natural harmonics, some musicians use natural tunings to write, perform, and record music. To some listeners this music sounds out of tune. To others it sounds pure and free of technological intervention, the way nature intended it to sound.

TEACHER NOTES

FOLLOW-UP ACTIVITIES

Textbook assignments

Assign textbook application problems on the geometric sequence and related formulas.

Writing prompts

- What did you learn in this activity?
- Do you think it matters that the frequency ratios are irrational numbers in even temperament?
- Many people debate the role of technology in our lives. Is using mathematics to create even temperament an application of technology? Can you think of any other example where the use of technology or mathematics to solve a problem might be controversial?

Extensions

- Use a tone generator to make recordings of Pythagorean tuning and even-tempered tuning. Present the recordings to the class and evaluate which sounds better.
- Find a way to record or perform a common melody or chord progression using Pythagorean tuning and even temperament and compare their sounds.

ANSWERS

Problem with Pythagorean Tuning

1. 7
2. 12
3. Twelve fifths will cover exactly seven octaves on the keyboard.
4. Mathematically, both methods should yield the same ratio for the big interval.
5. The ratio of a fifth is $\frac{3}{2}$. Stacking it 12 times means multiplying $\frac{3}{2}$ by itself 12 times, or $\frac{531,441}{4,096}$, which is about 129.746.
6. The ratio of an octave is $\frac{2}{1}$. Stacking it seven times means multiplying 2 by itself seven times, or $2^7 = 128$.
7. No; they are different by 1.746.

Even-Tempered Scale

1. Geometric sequence
2a. Twice as big
2b. Solution: Recall the formula for the nth term of a geometric sequence in the form taught in many textbooks:

$$t_n = t_1 r^{(n-1)}$$

In our case we are concerned with the distance of 12 intervals that, counting the first tone, comprise a sequence of 13 tones, so $n = 13$. Also, in the case of an octave, $t_{13} = 2t_1$, so substituting yields $2t_1 = t_1 r^{(13-1)}$.
Simplifying:

$$2t_1 = t_1 r^{12}$$
$$2 = r^{12}$$
$$r = 2^{1/12} = \sqrt[12]{2}$$

2c. Using this r-value we can substitute into the formula for the nth term of a geometric sequence to get a formula for the frequency of n half steps above the starting frequency:

$$f_n = f_1\left(2^{(n-1)/12}\right)$$

3. All the ratios are different:

Interval	Pythagorean	Term of series	Even-tempered
Second	$\frac{9}{8} = 1.125$	3	$2^{2/12} = 1.122$
Third	$\frac{81}{64} = 1.266$	5	$2^{4/12} = 1.260$
Fourth	$\frac{4}{3} = 1.333$	6	$2^{5/12} = 1.335$
Fifth	$\frac{3}{2} = 1.5$	8	$2^{7/12} = 1.498$
Sixth	$\frac{27}{16} = 1.688$	10	$2^{9/12} = 1.682$
Seventh	$\frac{243}{128} = 1.898$	12	$2^{11/12} = 1.888$

TEACHER NOTES

NATURAL HARMONICS

Musical Intervals and Frequency Ratios

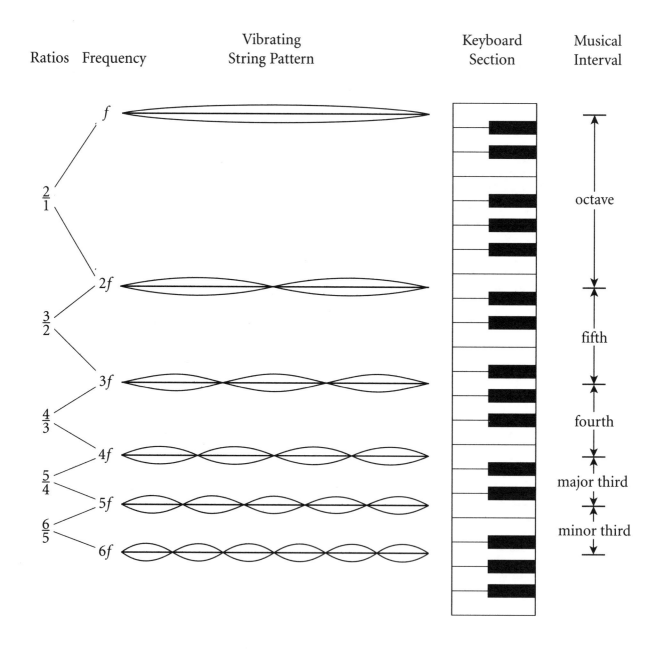

Ratios	Frequency	Vibrating String Pattern	Keyboard Section	Musical Interval

PROBLEM WITH PYTHAGOREAN TUNING

You have learned that the frequencies of the natural notes and the musical intervals they create determine the frequency values for the notes of the major scale. In this exploration, you will discover that this method has a problem.

1. The smallest musical interval in Western music is called a *half step*. It is the distance between any two adjacent keys, white or black. How many half steps (black and white keys) make a fifth?

2. How many half steps make an octave?

3. If you were to lay fifths end on end (sharing the adjoining note) and octaves end on end (sharing the adjoining note), how many fifths would it take before you covered an exact number of octaves? (*Note:* You can count up the keyboard or use some mathematical tools to make your job easier. If you're counting, be careful: Count a half step as two adjacent notes, whether they are black or white, and count the appropriate number of half steps for a fifth or an octave.) Explain how you got your answer, or show your work.

4. In question 3, the large interval made by exactly overlapping fifths and octaves nearly spanned the whole keyboard. Recall the frequency ratios for a fifth $\left(\frac{3}{2}\right)$ and an octave $\left(\frac{2}{1}\right)$. Remember that to find the frequency ratio of the resulting interval when intervals are laid end on end, the frequencies of the two smaller intervals are multiplied. Using this system we could calculate the frequency ratio for the large interval covered in question 3 using the ratio for the fifth. Since that interval is also an exact number of octaves, we could calculate it using octaves as well. Do you think both methods should give the same frequency ratio for the large interval?

5. Use the number of fifths that created the large interval in question 3 and the ratio for fifths to calculate the frequency ratio of the large interval.

6. Use the number of octaves that created the large interval in question 3 and the ratio for octaves to calculate the frequency ratio of the large interval.

7. Are your answers to questions 5 and 6 the same? If not, how far apart are they?

EVEN-TEMPERED SCALE

Your discovery in the Problem with Pythagorean Tuning worksheet does not represent an error in our method or some miscalculation. In fact, no successive number of pure fifths can ever equal an exact number of octaves. The question now arises: How does the piano get tuned when these intervals don't add up? On the piano, 12 successive fifths do equal 7 octaves! It appears that Pythagorean tuning doesn't work perfectly on the piano.

Musicians and mathematicians have used many different systems to resolve the problem. In 1636, French mathematician Marin Mersenne developed the simplest and most versatile solution, *even temperament*. It is used to tune instruments today. Let's discover it for ourselves.

1. Notice that on the piano there are actually 12 half-step intervals in an octave. Yet Pythagorean tuning yields different ratios for different half steps that occur at different places in the scale. We can fix the scale by setting it up so that the ratio is the same for all half steps throughout the scale. Recall various types of number sequences that you have studied. What type of number sequence has a common ratio between successive terms?

2. Think of the frequencies for the notes on the piano as that type of number sequence. Since there are 12 half steps in an octave, there are actually 13 notes in this sequence if you count the first and last notes.

    a. In this sequence, how much bigger than the first note is the thirteenth note?

    b. What is the common ratio between each term of the sequence?

    c. Use the ratio from 2b to find a formula you can use to determine the frequency of any tone from a given tone. (*Hint:* Use the formula for the *n*th term in a geometric sequence.)

3. Using this formula, calculate the ratios for the second, third, fourth, fifth, sixth, and seventh intervals. Compare your values to the ratios found in Part I using Pythagorean tuning. Which values are different?